Materials Science and Fuel Technologies of Uranium and Plutonium Mixed Oxide

Materials Science and Fuel Technologies of Uranium and Plutonium Mixed Oxide offers a deep understanding of mixed oxide (MOX) properties for nuclear fuels that will be useful for performance evaluation. It also reviews fuel property simulation technology and an irradiation behavior model required for performance evaluation.

Based on research findings, the book investigates various physical property data in order to develop MOX fuel for sodium-cooled fast reactors. It discusses a database of MOX properties, including oxygen potential, melting temperature, lattice parameter, sound speeds, thermal expansion, thermal diffusivity, oxygen self-diffusion, and chemical diffusion coefficients, that was used to derive a science-based model of MOX properties (Sci-M Pro) for fuel-performance code development.

Features:

- Concisely covers the essential aspects of MOX nuclear fuels.
- Explores MOX nuclear fuels by systematically evaluating various physical property values using a behavior model.
- Presents fuel property simulation technology.
- Considers oxygen potential, lattice parameter, sound speeds, and oxygen self-diffusion.
- Discusses melting temperature, thermal expansion, thermal diffusivity, and chemical diffusion coefficients.

The book will be useful for researchers and engineers working in the field of nuclear fuels and nuclear materials.

Dr. Masato Kato is currently senior principal researcher at the Sector of Fast Reactor Research and Development at the Japan Atomic Energy Agency. He received his doctorate from Tohoku University in 2009. His work focuses specifically on the physical properties and nuclear fuel technology of plutonium oxide fuels. He is a member of the Atomic Energy Society of Japan and is the President of its Nuclear Fuel subcommittee in 2021–2022.

Dr. Masahiko Machida is currently the Deputy Director at Computational Science and E-systems at Japan Atomic Energy Agency. He received his doctorate from Tohoku University in 2000. His research area is theoretical and computational physics and chemistry on various materials, including nuclear fuel compounds. He is a member of the Atomic Energy Society of Japan and Physical Society of Japan.

Materials Science and Fuel Technologies of Uranium and Plutonium Mixed Oxide

Edited by
Masato Kato and Masahiko Machida

CRC Press
Taylor & Francis Group
Boca Raton London New York

CRC Press is an imprint of the
Taylor & Francis Group, an **informa** business

First edition published 2023
by CRC Press
6000 Broken Sound Parkway NW, Suite 300, Boca Raton, FL 33487-2742

and by CRC Press
4 Park Square, Milton Park, Abingdon, Oxon, OX14 4RN

CRC Press is an imprint of Taylor & Francis Group, LLC

ISBN: 978-1-032-28713-3 (hbk)
ISBN: 978-1-032-28717-1 (pbk)
ISBN: 978-1-003-29820-5 (ebk)

DOI: 10.1201/9781003298205

Typeset in Times
by codeMantra

Contents

Preface

We are now facing the crucial problem of global warming. One way to address this is to develop energy technologies reducing greenhouse gas emissions. Nuclear energy is a candidate for a sustainable energy source for the future as it does not emit CO_2 gas. In order to use plutonium as a nuclear fuel, it is necessary to realize a nuclear fuel cycle. And, it will be possible to reduce the use of fossil fuels used for thermal power generation over a long period of time, which will contribute to a sustainable environment.

Innovative and advanced nuclear reactors using plutonium fuel have been developed. Irradiation tests are indispensable to develop new nuclear fuel, and nuclear fuel performance and safety must be demonstrated. A technology that accurately simulates irradiation behavior complimenting the irradiation test can significantly reduce the cost, time, and labor involved in nuclear fuel studies. Safety and reliability can be significantly improved through nuclear fuel irradiation behavior simulation. To evaluate the performance of nuclear fuel, it is necessary to know the physical and chemical properties of the fuel at high temperatures and develop a behavior model describing various phenomena occurring during irradiation. In previous studies, empirical methods with fitting parameters have been used in many parts of model development; however, empirical techniques can give different results in areas with no data. Therefore, this study constructs a scientific descriptive model to extrapolate the basic characteristics of fuel to the composition and temperature and develops an irradiation behavior analysis code to which the model is applied.

Our research group has measured various basic physical property data for uranium–plutonium mixed oxides (MOX) and created the latest database. Furthermore, we applied first-principles and molecular dynamics calculations to deepen our understanding of the mechanism and derived a consistent science-based physical property model for all physical property data. We are also developing a behavior model to incorporate this model into the irradiation behavior analysis code. Specifically, it will be possible to accurately reproduce the effects of Pu and oxygen contents on the properties and behavior. This book summarizes the scientific findings since 2005 on the basic properties of MOX fuel and the application of fuel technology. The basic properties and nuclear fuel technology will contribute to plutonium fuel development for future energy security.

Masato Kato

Contributors

Shun Hirooka
Plutonium Fuel Development Center
Japan Atomic Energy Agency
Tokai-mura, Japan

Yoshihisa Ikusawa
Plutonium Fuel Development Center
Japan Atomic Energy Agency
Tokai-mura, Japan

Keita Kobayashi
Center for Computational Science and
   e-Systems
Japan Atomic Energy Agency
Kashiwa, Japan

Koji Maeda
Fuels and Materials Department
Japan Atomic Energy Agency
Oarai-machi, Japan

Taku Matsumoto
Plutonium Fuel Development Center
Japan Atomic Energy Agency
Tokai-mura, Japan

Kyoichi Morimoto
Plutonium Fuel Development Center
Japan Atomic Energy Agency
Tokai-mura, Japan

Shinya Nakamichi
Plutonium Fuel Development Center
Japan Atomic Energy Agency
Tokai-mura, Japan

Hiroki Nakamura
Center for Computational Science and
   e-Systems
Japan Atomic Energy Agency
Kashiwa, Japan

Takayuki Ozawa
Fuel Cycle Design Office
Japan Atomic Energy Agency
Oarai-machi, Japan

Shinji Sasaki
Fuels and Materials Department
Japan Atomic Energy Agency
Oarai-machi, Japan

Masashi Watanabe
Fuel Cycle Design Office
Japan Atomic Energy Agency
Oarai-machi, Japan

Keisuke Yokoyama
Fuels and Materials Department
Japan Atomic Energy Agency
Oarai-machi, Japan

1 Introduction

Uranium (U) and plutonium (Pu) constitute nuclear fuels and generate enormous energy. U is a naturally occurring element, but Pu is produced by a nuclear reaction during nuclear fuel burning. Therefore, Pu can be used as nuclear fuel by extracting it from spent fuel in the reprocessing process and processing it as fuel and using it as energy. Research on Pu fuel has developed fuels in various chemical forms, such as oxides, metals, and nitrides, as nuclear fuels for sodium-cooled fast reactors. U–Pu mixed oxides (MOX) are currently used in commercial light-water reactors. Furthermore, Pu-containing oxide fuels can be used in various nuclear, light-water, heavy-water, gas-cooled, and sodium-cooled fast reactors to secure future energy [1–4].

The reprocessing process required to use Pu produces high-level waste containing minor actinide (MA) elements, such as Am and Np, which are long-lived radioactive materials. Because these elements can be burned in fast reactors, research and development have been conducted on MA-containing fuels for the future nuclear fuel cycle. Separating MA elements from spent fuels in the reprocessing process and recycling them as fuel can reduce the volume and toxicity of high-level waste. Fuel can burn without increasing MA by confining MA in the nuclear fuel cycle and recycling it as an MA-bearing fuel in a fast reactor. Realizing this technology can significantly reduce the burden on the environment for high-level waste treatment [2,3,5,6].

To burn in a nuclear reactor as nuclear fuel, it is necessary to guarantee its integrity during steady operations and transition conditions and efficiently obtain it as thermal energy. Therefore, it is essential to know the thermal and mechanical properties and evaluate the fuel performance in fuel development, and efforts have been made to measure these properties. However, property measurements of Pu-containing fuels were challenging, and the data have considerable uncertainty. Pu materials must be handled inside a glove box, and facilities are limited. Furthermore, the chemical stability of the oxide causes high-temperature measurement challenges. Actinide oxides are nonstoichiometric compounds, and the oxygen content in oxides changes depending on their temperature, composition, and atmosphere. The challenge of determining oxygen content increases the uncertainty of various properties and requires a large safety margin in fuel performance analysis.

It is necessary to understand and describe various phenomena occurring during irradiation in nuclear fuel developments. Numerous physical property data are required for fuel performance evaluation. To apply the physical property data to fuel performance evaluation, it is indispensable to understand the mechanism consistently to ensure the high reliability of various physical property data. Nonstoichiometric actinide oxides can exist stably as fluorite structures with various oxygen-to-metal (O/M) ratios. The O/M ratio strongly affects the thermophysical properties. One experimental challenge of MOX is determining the O/M ratio. Through potential oxygen measurement in previous studies, the authors have developed O/M ratio control technology and measured the physical properties. Furthermore, it is vital

DOI: 10.1201/9781003298205-1

to understand the influence of $5f$ electrons on actinide oxides' physical properties and their mechanisms. The $5f$ electrons affect the lattice defects' electronic state and behavior and are closely related to the mechanism of physical characteristic data. It is challenging to understand such a mechanism by experiment alone, and applying computational science is indispensable. We reviewed studies on computational science and applied first-principles and molecular dynamics calculations to advance our understanding of the mechanism.

The authors' research group has conducted measurements as functions of the Pu content parameters, MA content, O/M ratio, and temperature [7–44]. All data measured were listed and reviewed, more than half of all MOX data globally. An integrated MOX property science-based model was derived and applied to model development to analyze the irradiation behavior of Pu fuels of various compositions. The parameters and symbols used in this study are explained in the Notation table.

In this book, we evaluated the basic characteristics of MOX. We proceeded with modeling to explain all basic characteristics consistently, with an awareness of application to the irradiation behavior analysis code. Therefore, MOX-fuel behavior during a severe accident and waste behavior during long-time storage are excluded. Insufficient physical properties and models still exist, such as the combustion effect, but I hope they will be updated based on this document.

REFERENCES

1. Fridman, E., et al., Axial discontinuity factors for the nodal diffusion analysis of high conversion BWR cores. *Annals of Nuclear Energy*, 2013. **62**: pp. 129–136.
2. Kato, M., et al., 2.01: Uranium oxide and MOX production, in Comprehensive Nuclear Materials (Second Edition). Editor-in-Chief: R. J.M. Konings and R. Stoller, 2020, Amsterdam, Netherlands, Elsevier Ltd..
3. Dimayuga, F.C., CANDU MOX fuel fabrication development. *Transactions of the American Nuclear Society, MOX Fuel Development and Testing Experience—I*, 1997. **77**: pp. 150–152.
4. Provost, J.L. and M. Debes, MOX and UOX PWR fuel performances EDF operating experience. *Journal of Nuclear Science and Technology*, 2006. **43**(9): pp. 960–962
5. Pillon, S., Actinide-bearing fuels and transmutation targets, in *Comprehensive Nuclear Materials*, R. J. M. Koning, Ed. 2012. Oxford, Elsevier.
6. McClellan, K., et al., *Summary of the Minor Actinide-bearing MOX AFC-2C and -2D Irradiations*. 2013: IAEA; Vienna (International Atomic Energy Agency (IAEA)); International Atomic Energy Agency, Nuclear Power Technology Development Section and Nuclear Fuel Cycle and Materials Section, Vienna (Austria); French Alternative Energies and Atomic Energy Commission (CEA), Gif-sur-Yvette Cedex (France); French Nuclear Energy Society (SFEN), Paris (France). Medium: X; Size: 27 page(s).
7. Kato, M., et al., Oxygen potentials of plutonium and uranium mixed oxide. *Journal of Nuclear Materials*, 2005. **344**(1–3): pp. 235–239.
8. Kato, M., T. Tamura, and K. Konashi, Oxygen potentials of mixed oxide fuels for fast reactors. *Journal of Nuclear Materials*, 2009. **385**(2): pp. 419–423.
9. Kato, M., K. Konashi, and N. Nakae, Analysis of oxygen potential of $(U_{0.7}Pu_{0.3})O_{2\pm x}$ and $(U_{0.8}Pu_{0.2})O_{2\pm x}$ based on point defect chemistry. *Journal of Nuclear Materials*, 2009. **389**(1): pp. 164–169.
10. Nakamichi, S., M. Kato, and T. Tamura, Influences of Am and Np on oxygen potentials of MOX fuels. *Calphad*, 2011. **35**(4): pp. 648–651.

11. Kato, M., et al., Oxygen potential of $(U_{0.88}Pu_{0.12})O_{2\pm x}$ and $(U_{0.7}Pu_{0.3})O_{2\pm x}$ at high temperatures of 1673–1873K. *Journal of Nuclear Materials*, 2011. **414**(2): pp. 120–125.

12. Kato, M., et al., Measurement of oxygen potential of $(U_{0.8}Pu_{0.2})O_{2\pm x}$ at 1773 and 1873 K, and its analysis based on point defect chemistry. *Calphad*, 2011. **35**(4): pp. 623–626.

13. Komeno, A., et al., Oxygen potentials of PuO_{2-x}. *MRS Proceedings*, 2012. **1444**: pp. 58–89.

14. Matsumoto, T., et al., Oxygen potential measurement of $(Pu_{0.928}Am_{0.072})O_{2-x}$ at high temperatures. *Journal of Nuclear Science and Technology*, 2014. **52**(10): pp. 1296–1302.

15. Kato, M., et al., Oxygen potentials, oxygen diffusion coefficients and defect equilibria of nonstoichiometric $(U, Pu)O_{2\pm x}$. *Journal of Nuclear Materials*, 2017. **487**: pp. 424–432.

16. Hirooka, S., et al., Oxygen potential measurement of $(U, Pu, Am)O_{2\pm x}$ and $(U, Pu, Am, Np)O_{2\pm x}$. *Journal of Nuclear Materials*, 2020. **542**: p. 152424.

17. Kato, M., Oxygen potentials and defect chemistry in nonstoichiometric $(U, Pu) O_2$, in *Stoichiometry and Materials Science - When Numbers Matter*, A. Innocenti and N. Kamarulzaman, Eds. 2012. Intech., London, UK

18. Hirooka, S., et al., Relative oxygen potential measurements of $(U, Pu)O_2$ with $Pu = 0.45$ and 0.68 and related defect formation energy. *Journal of Nuclear Materials*, 2022. **558**: p. 153375.

19. Watanabe, M., M. Kato, T. Sunaoshi, Oxygen potential measurement and point defect chemistry of UO_2. *Transactions of the American Nuclear Society*, June 12–16, 2016. **114**: pp. 1081–1082. New Orleans, Louisiana.

20. Kato, M., et al., Solidus and liquidus of plutonium and uranium mixed oxide. *Journal of Alloys and Compounds*, 2008. **452**(1): pp. 48–53.

21. Kato, M., et al., Solidus and liquidus temperatures in the UO_2–PuO_2 system. *Journal of Nuclear Materials*, 2008. **373**(1–3): pp. 237–245.

22. Kato, M., et al., The effect of oxygen-to-metal ratio on melting temperature of uranium and plutonium mixed oxide fuel for fast reactor. *Transactions of the Atomic Energy Society of Japan*, 2008. **7**(4): pp. 420–428.

23. Kato, M. Melting temperatures of oxide fuel for fast reactors, in *International Congress on Advances in Nuclear Power Plants 2009*. 2009. Tokyo, Curran Associates, Inc., Atomic Energy Society of Japan (AESJ).

24. Kato, M. and K. Konashi. Lattice parameters of $(U, Pu, Am, Np)O_{2-x}$, in *Recent Advances in Actinide Science*, edited by Iain May, Rebecca Alvares and Nicholas BryanRSC Publishing, c2006, in *Actinides 2005*. 2005. Manchester, UK

25. Kato, M. and K. Konashi, Lattice parameters of $(U, Pu, Am, Np)O_{2-x}$. *Journal of Nuclear Materials*, 2009. **385**(1): pp. 117–121.

26. Kato, M., et al., Self-radiation damage in plutonium and uranium mixed dioxide. *Journal of Nuclear Materials*, 2009. **393**(1): pp. 134–140.

27. Hirooka, S. and M. Kato, Sound speeds in and mechanical properties of $(U, Pu)O_{2-x}$. *Journal of Nuclear Science and Technology*, 2017. **55**(3): pp. 356–362.

28. Uchida, T., et al., Thermal properties of UO_2 by molecular dynamics simulation. *Progress in Nuclear Science and Technology*, 2011. **2**: pp. 298–602.

29. Uchida, T., et al., Thermal expansion of PuO_2. *Journal of Nuclear Materials*, 2014. **452**(1–3): pp. 281–284.

30. Kato, M., et al., Thermal expansion measurement and heat capacity evaluation of hypostoichiometric $PuO_{2.00}$. *Journal of Nuclear Materials*, 2014. **451**(1–3): pp. 78–81.

31. Kato, M., et al., Thermal expansion measurement of $(U, Pu)O_{2-x}$ in oxygen partial pressure-controlled atmosphere. *Journal of Nuclear Materials*, 2016. **469**: pp. 223–227.

32. Morimoto, K., et al. The influence of Pu-content on thermal conductivities of $(U, Pu)O_2$ solid solution, in Proceedings of International Conference on Fast Reactors and Related Fuel Cycles *2009*. 2009. Kyoto, Proceedings Series - International Atomic Energy Agency.

33. Morimoto, K., et al., Recovery behaviours of thermal conductivities in self-irradiated MOX fuel, in *IOP Conference Series: Materials Science and Engineering*, 2010. **9**: p. 012008. Bristol, UK, IOP Publishing.

34. Morimoto, K., M. Kato, and M. Ogasawara, Thermal diffusivity measurement of $(U, Pu)O_{2-x}$ at high temperatures up to 2190K. *Journal of Nuclear Materials*, 2013. **443**(1–3): pp. 286–290.

35. Morimoto, K., et al., Thermal conductivities of hypostoichiometric $(U, Pu, Am)O_{2-x}$ oxide. *Journal of Nuclear Materials*, 2008. **374**(3): pp. 378–385.

36. Morimoto, K., et al., Thermal conductivity of $(U, Pu, Np)O_2$ solid solutions. *Journal of Nuclear Materials*, 2009. **389**(1): pp. 179–185.

37. Morimoto, K., et al., Thermal conductivities of $(U, Pu, Am)O_2$ solid solutions. *Journal of Alloys and Compounds*, 2008. **452**(1): pp. 54–60.

38. Matsumoto, T., et al., Thermal conductivity measurement of $(Pu_{1-x}Am_x)O_2$ (x=0.028, 0.072). *Journal of Alloys and Compounds*, 2015. **629**: pp. 92–97.

39. Yokoyama, K., et al., Measurements of thermal conductivity for near stoichiometric $(U_{0.7-z}Pu_{0.3}Am_z)O_2$ (z = 0.05, 0.10, and 0.15). *Nuclear Materials and Energy*, 2022. **31**: p. 101156.

40. Kato, M., et al., Physical properties and irradiation behavior analysis of Np- and Am-bearing MOX fuels. *Journal of Nuclear Science and Technology*, 2011. **48**: pp. 646–653.

41. Kato, M., et al., Oxygen chemical diffusion in hypo-stoichiometric MOX. *Journal of Nuclear Materials*, 2009. **389**(3): pp. 416–419.

42. Kato, M., T. Uchida, and T. Sunaoshi, Measurement of oxygen chemical diffusion in PuO_{2-x} and analysis of oxygen diffusion in PuO_{2-x} and $(Pu, U)O_{2-x}$. *Physica Status Solidi (c)*, 2013. **10**(2): pp. 189–192.

43. Watanabe, M., M. Kato, and T. Sunaoshi, Oxygen self-diffusion in near stoichiometric $(U, Pu)O_2$ at high temperatures of 1673–1873 K. *Journal of Nuclear Materials*, 2020. **542**: p. 152472.

44. Watanabe, M., T. Sunaoshi, and M. Kato, Oxygen chemical diffusion coefficients of $(U, Pu)O_{2-x}$. *Defect and Diffusion Forum*, 2017. **375**: pp. 84–90.

2 Experimental Techniques

Masato Kato
Japan Atomic Energy Agency

CONTENTS

2.1 SAMPLE PREPARATION

Masato Kato and Shinya Nakamichi

Because $(U, Pu)O_2$ pellet preparation is conducted using powder metallurgy, it is essential to obtain composition homogeneity and pore distribution in property measurements [1–38]. As the starting material for sample preparation in physical property measurements, the coprecipitation method obtained excellent homogeneity in previous works. Our research group employed the microwave-heating denitration method, where the Pu and U compositions were adjusted in nitrate solution and heated into powder [39]. Some minor actinide (MA)-bearing mixed oxides (MOX) samples were prepared using the mechanical blending method; however, some problems were induced in the sample preparation.

One problem is forming Pu spots in the MOX pellets. Pellet sintering was conducted at approximately 2,000 K. Considering actinide elements' diffusion coefficient, the U and Pu diffusion distance is several micrometers, even in 100 h of heat treatment. Heating treatments cannot homogenize heterogeneous spots of several hundreds of micrometers formed in the sintered pellets. Therefore, a homogeneous sample using a mechanical blending method can be obtained by paying sufficient attention to the properties of the raw material powder, mixing methods, and initial sintering conditions.

Sintering behavior of MOX pellets was investigated using dilatometry [40,41]. MOX pellet sintering proceeds at 1,500–2,000 K in general conditions of Ar–5% H_2 mixing gas. The sintering temperature range shifts to as low as approximately 200 K when the oxygen partial pressure in the sintering atmosphere is adjusted, and sintering is performed at around O/M ratio = 2.00. In previous studies, the apparent activation energy in sintering was evaluated to be 919 and 482 kJ/mol, respectively, for the sintering of O/M ratio = 2.00 and low O/M. It was shown that the sintering mechanisms differed between both conditions [40].

Figure 2.1 shows element-mapping analysis results measured using an electron probe microanalyzer on the cross section of MA-bearing MOX pellets prepared

DOI: 10.1201/9781003298205-2

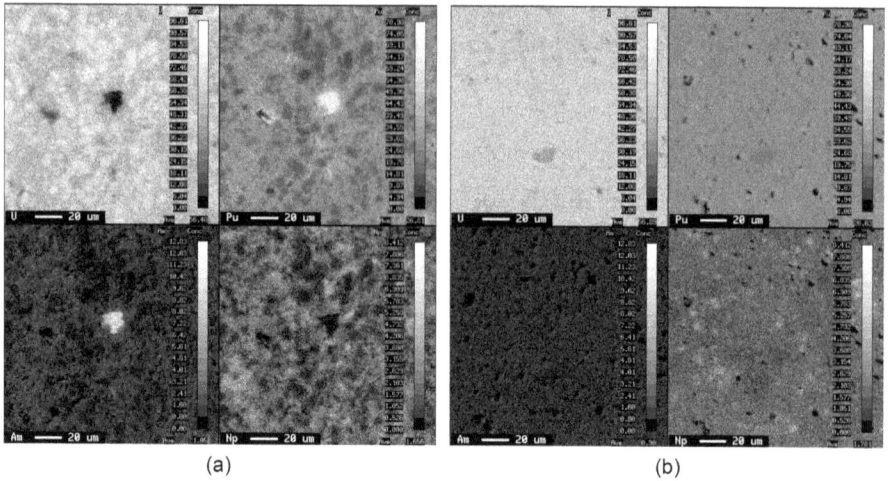

FIGURE 2.1 Element-mapping analysis of minor actinide (MA)-bearing mixed oxide (MOX) pellets with (a) low O/M and (b) O/M=2.00 which were sintered at 1,973 K.

using a mechanical blending method in two conditions. MOX pellets sintered at O/M ratio = 2.00 were well homogeneous. Takeuchi et al. [41] reported that grain boundary diffusion is a crucial mechanism in initial sintering to obtain homogeneous pellets. Nakamichi et al. [40] explained the dependence of sintering behavior on the O/M ratio using UO_2–UO_2 and UO_2–PuO_2 diffusion mechanisms. It is needed to deepen the understanding of sintering mechanisms.

The sample was prepared with density as a parameter in thermal conductivity studies [8]. Organic additives, such as zinc stearate and cellulose, were used to control the pellet density. The additives were removed in the presintering process and sintered into pellets. The pellet density decreased with the additive amount, and MOX samples of 95%–84%TD were prepared and used in thermal diffusivity measurements. Mixing and removing additives form heterogeneous pores and cracks under certain conditions, unexpectedly decreasing thermal conductivity; therefore, the microstructure of the sample cross-section must be observed before the measurement [14].

It is vital to control the grain size and density during sample preparation. Crystal grain growth also changes significantly depending on the O/M ratio. Figure 2.2 (a) and (b) shows the microstructure of the cross section of MOX pellets with O/M ratio, which is equal to 2.00 and 1.99, respectively. The heat treatment was conducted in a P_{O_2}-controlled atmosphere to maintain a constant O/M ratio.

In the heat treatment at 1,923 K for 5 h for O/M ratio = 2.00, the grain growth rate was high, and a grain size of approximately 25 μm was obtained. However, when heated at a low O/M ratio (1.99), the grain size was approximately 15 μm. Nakamich et al. [40] evaluated grain growth rate using Eq. (2.1).

$$G^3 - G_0^{\ 3} = kt \qquad\qquad (2.1)$$

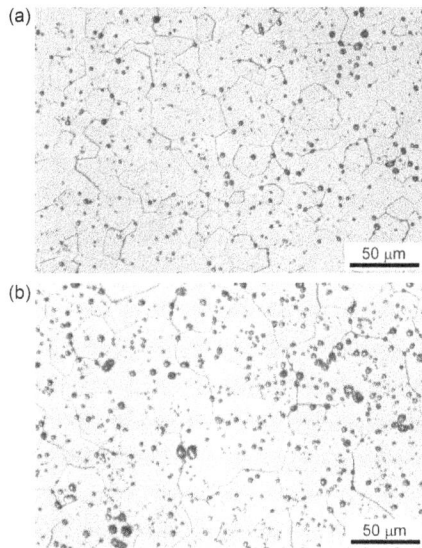

FIGURE 2.2 Microstructure on the cross section of pellets with (a) O/M=1.99 and (b) O/M=2.00 sintered at 1,923 K for 5 h.

They reported that k was 1,143 and 2,348 μm^3/h for pellets with O/M ratios of 1.99 and 2.00, respectively, at 1,973 K. The grain size might affect fuels' physical characteristics and irradiation behavior, such as affecting the microstructure change. Thus, the preparation method for the raw material powder and the heat treatment technique for obtaining sintered pellets with a homogeneous microstructure are needed.

The experiment also has radiation concerns to consider in Pu fuel experiments. Pu is an emitter element, so it must be prevented from being taken into the body to avoid internal exposure. Reactor-grade Pu also contains a small percentage of Am-241 produced by the beta decay of Pu-241. Gamma rays from Am-241 are problematic due to external exposure during the experiment. Therefore, every experiment was performed in a glove box with lead-containing glass for gamma-ray shielding. Every experimental equipment was installed in a glove box controlled by negative pressure, and electricity, signals, gas, and cooling water connections were sealed and connected. Long-term use causes various problems, such as equipment failure and component wear, but we have designed and experimented with equipment for long-term maintenance. Researchers and technicians have made more efforts to maintain the facilities and equipment and obtain data that are more experimental.

2.2   O/M CONTROL

Masato Kato and Masashi Watanabe

Two O/M control techniques were applied in the property measurements: heat treatment and in situ P_{O_2}-controlling methods. Figure 2.3 shows the relationship of O/M–T

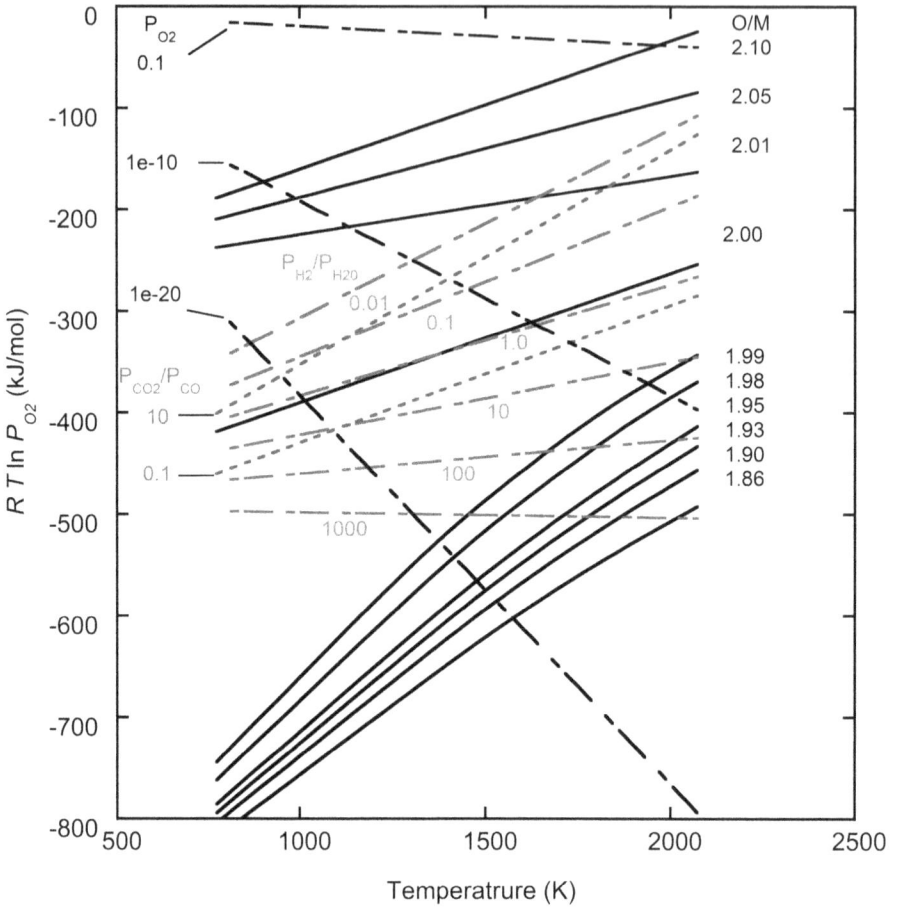

FIGURE 2.3 Relationship of O/M–T–$\Delta \bar{G}_{O_2}$.

$\Delta \bar{G}_{O_2}$. In the heat treatment method, the sintered pellets were heated at 1,123 K at an atmosphere of $\Delta \bar{G}_{O_2} = 380$ kJ/mol to adjust to O/M ratio = 2.00, and the second heat treatment was conducted to control the O/M ratio. The second heat treatment's condition was determined from Figure 2.3, appropriate for obtaining the required O/M ratio. For example, the MOX sample with O/M ratio = 1.95 could be obtained from the sintering at 2,000 K in −430 kJ/mol atmosphere. The heat treatment was conducted until P_{O_2} at the furnace's outlet attained equilibrium. The O/M ratio was determined from the weight change in the sample in the second heat treatment using Eq. (2.2).

$$\frac{O}{M} = 2 + \frac{M}{16} \cdot \frac{\Delta W}{W}, \qquad (2.2)$$

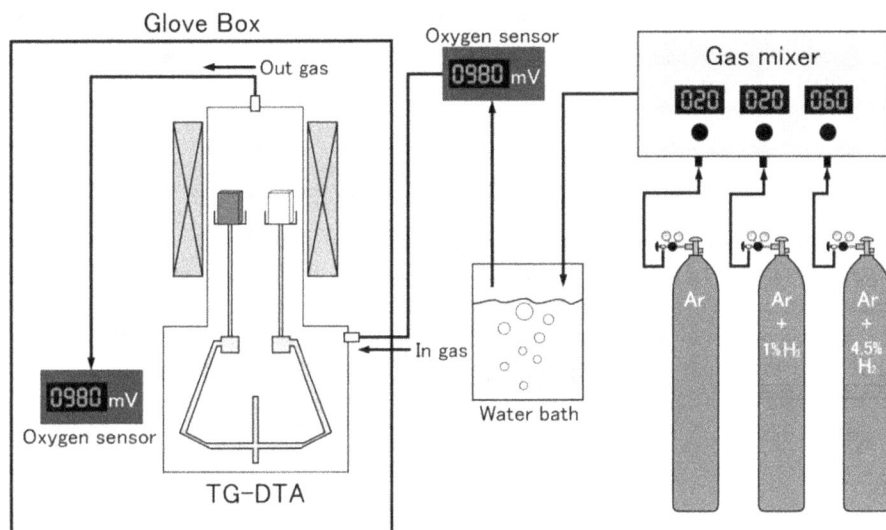

FIGURE 2.4 Gas system for controlling oxygen partial pressure.

The samples obtained using this method were subjected to lattice parameters, melting temperature, sound speed, and thermal diffusivity measurements. Thermal diffusivity measurements of MOX samples with various O/M ratios were conducted in a high vacuum atmosphere. Samples with O/M ratio = 2.00 could be measured up to 1,700 K, but the samples were reduced at higher temperatures. The samples' weight measurements before and after the experiments were essential to confirm no O/M ratio change.

To control the O/M ratio, H_2-content gas-added H_2O, CO/CO_2 mixed gas, and an oxygen pump were used. Figure 2.3 shows adjustable ranges using the three methods. In our study, H_2-content gas was used for measuring hypostoichiometric MOX. Figure 2.4 shows the gas-controlling system used in previous studies. Four pure 0.005%H_2, 0.5%H_2, and 5%H_2–Ar mixed gases were prepared for the experiments. A mass flow controller can adjust the hydrogen content using those gases. A diffusion-type humidifier, which can add moisture from 10 ppm to thousands of ppm, added moisture to the gas. Moisture of more than 100 ppm was needed to control the P_{O_2} stably. The ratio of P_{H_2} / P_{H_2O} was adjusted during the measurements at high temperatures to measure the thermal expansion, oxygen diffusion coefficient, and oxygen potential. A stabilized zirconia oxygen sensor was used to monitor Po_2 at the inlet and outlet of the measuring device.

The P_{O_2} equilibrium reaction of $H_2O = H_2 + 1/2O_2$ controlled the P_{O_2} in the atmosphere, and the oxygen sensor measured P_{O_2} at 973 K. The gas-phase equilibrium was correlated to the standard Gibbs free energy of water formation, ΔG_f using the following equations:

$$\Delta G_f = RT \ln \frac{P_{H_2O}}{P_{H_2} P_{O_2}^{1/2}} \qquad (2.3)$$

$$\Delta G_f = -246440 + 54.8\,T \tag{2.4}$$

The electromotive force of the oxygen sensor was obtained using Eq. (2.5).

$$E = \frac{RT}{4F}\ln\frac{\left(P_{O_2}\right)_A}{\left(P_{O_2}\right)_B}. \tag{2.5}$$

$\left(P_{O_2}\right)_B$ equals 0.206 when air is used as the reference gas. The $\left(P_{O_2}\right)_A$ in gas for the experiments can be calculated. We can obtain the P_{H_2O}/P_{H_2} ratio at a measurement temperature of $\left(P_{O_2}\right)_A$ by inserting the measured $\left(P_{O_2}\right)_A$ into Eq. (2.3).

The P_{O_2} values at the measurement temperatures were determined from Eqs. (2.3) and (2.4), assuming that the P_{H_2O}/P_{H_2} ratio obtained from the oxygen sensor's data is maintained constant, even at the experimental temperature. The $\Delta\bar{G}_{O_2}$ is described using

$$\Delta\bar{G}_{O_2} = RT\ln P_{O_2} \tag{2.6}$$

The uncertainty of $\Delta\bar{G}_{O_2}$ is determined to be ±10 kJ/mol from the difference of P_{O_2} between the inlet and outlet gases.

For the CO_2/CO gas, Eqs. (2.7) and (2.8) are used instead of Eqs. (2.3) and (2.4).

$$RT\ln\frac{P_{CO_2}}{P_{CO}\cdot P_{O_2}^{1/2}} \tag{2.7}$$

$$\Delta G_f = -282{,}400 + 86.81\,T \tag{2.8}$$

The oxygen pump's principle is the same as the oxygen sensor, and the difference is to control $\left(P_{O_2}\right)_A$ by adjusting the voltage E. A gas chemical reaction is not used for P_{O_2} controlling.

From Figure 2.3, we can find that the gas of $P_{H_2O}/P_{H_2} = 1.0$ can control O/M ratio $= 2.00$ in the wide temperature range. It is challenging to adjust the low O/M ratio at temperatures lower than 1,500 K. The low O/M MOX sample was obtained from the high-temperature heat treatment. The property measurements were conducted in a high vacuum atmosphere, and the oxygen pump was employed for measuring hyperstoichiometric MOX. Various heat treatment conditions were used to adjust the O/M ratio and measure properties.

2.3　REACTION WITH HIGH-TEMPERATURE MATERIAL

Masato Kato

Metallic material, such as Mo, Ta, W, or Re, was used as capsule and sample holders in the high-temperature experiments, with melting temperatures of 2,896, 3,258, 3,695, and 3,459 K, respectively. It is crucial to select the material considering chemical stability at experimental conditions.

Molybdenum (Mo) is frequently used in experiments. Precautions for using Mo are its extremely high vapor pressure of Mo oxide, and the transformation from Mo

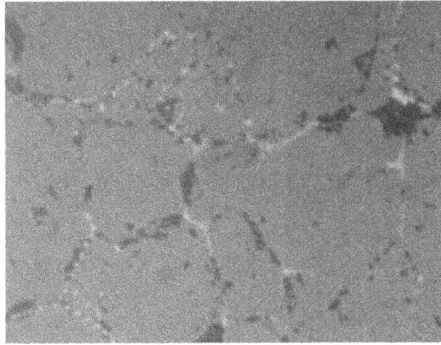

FIGURE 2.5   Reaction between $PuO_{1.9}$ and Re (white area) after heating at 2,900 K.

metal to Mo oxide causes rapid evaporation. The phase transformation occurs at around P_{O_2} of O/M ratio $= 2.00$ in MOX. Mo reacts with oxygen in (U, Pu)O_2 and becomes oxide vapor. Huge Mo material losses occur between Mo and MOX samples at the contact place. Therefore, Mo usage is restricted to the reducing atmosphere. Tantalum has large oxygen solubility, and reactions with hydrogen occur. Therefore, use in an atmosphere containing H_2 gas should be avoided at temperatures higher than 2,000 K.

Tungsten has been used at temperatures higher than the melting point of UO_2 in previous studies. In the melting temperature measurements, W was used as a capsule material. Valuable data were obtained in measuring 0%–20% Pu content MOX. In measuring high Pu content MOX of more than 30%, MOX reacted with the W capsule and could not measure an accurate melting temperature. In high Pu content measurements, Re inner was used. However, the reaction was observed in PuO_2 measurement, even when using the Re inner (Figure 2.5). The Institute for Transuranium Elements research group developed the laser method for PuO_2 measurement, which melts the MOX sample's local area, preventing it from reacting between high-temperature material and MOX.

REFERENCES

1. Yokoyama, K., et al., Measurements of thermal conductivity for near stoichiometric $(U_{0.7-z}Pu_{0.3}Am_z)O_2$ ($z = 0.05$, 0.10, and 0.15). *Nuclear Materials and Energy*, 2022. **31**: p. 101156.
2. Watanabe, M., T. Sunaoshi, and M. Kato, Oxygen chemical diffusion coefficients of (U, Pu)O_{2-x}. *Defect and Diffusion Forum*, 2017. **375**: p. 84–90.
3. Watanabe, M., M. Kato, and T. Sunaoshi, Oxygen self-diffusion in near stoichiometric (U, Pu)O_2 at high temperatures of 1673–1873 K. *Journal of Nuclear Materials*, 2020. **542**: p. 152472.
4. Watanabe, M., M. Kato, T. Sunaoshi, Oxygen potential measurement and point defect chemistry of UO_2. *Transactions of the American Nuclear Society*, June 12–16, 2016. **114**: 1081–1082. New Orleans, Louisiana.

5. Uchida, T., et al., Thermal expansion of PuO_2. *Journal of Nuclear Materials*, 2014. **452**(1–3): pp. 281–284.

6. Uchida, T., et al., Thermal properties of UO_2 by molecular dynamics simulation. *Progress in Nuclear Science and Technology*, 2011. **2**: pp. 298–602.

7. Nakamichi, S., M. Kato, and T. Tamura, Influences of Am and Np on oxygen potentials of MOX fuels. *Calphad*, 2011. **35**(4): pp. 648–651.

8. Morimoto, K., et al., Thermal conductivities of $(U, Pu, Am)O_2$ solid solutions. *Journal of Alloys and Compounds*, 2008. **452**(1): pp. 54–60.

9. Morimoto, K., et al., Thermal conductivity of $(U, Pu, Np)O_2$ solid solutions. *Journal of Nuclear Materials*, 2009. **389**(1): pp. 179–185.

10. Morimoto, K., et al., Thermal conductivities of hypostoichiometric $(U, Pu, Am)O_{2-x}$ oxide. *Journal of Nuclear Materials*, 2008. **374**(3): pp. 378–385.

11. Morimoto, K., M. Kato, and M. Ogasawara, Thermal diffusivity measurement of $(U, Pu)O_{2-x}$ at high temperatures up to 2190K. *Journal of Nuclear Materials*, 2013. **443**(1–3): pp. 286–290.

12. Morimoto, K., et al., Recovery behaviours of thermal conductivities in self-irradiated MOX fuel, in *IOP Conference Series: Materials Science and Engineering*, 2010. **9**: p. 012008. IOP Publishing, Bristol, UK

13. Matsumoto, T., et al., Oxygen potential measurement of $(Pu_{0.928}Am_{0.072})O_{2-x}$ at high temperatures. *Journal of Nuclear Science and Technology*, 2014. **52**(10): pp. 1296–1302.

14. Matsumoto, T., et al., Thermal conductivity measurement of $(Pu_{1-x}Am_x)O_2$ ($x = 0.028$, 0.072). *Journal of Alloys and Compounds*, 2015. **629**: pp. 92–97.

15. Kato, M. and K. Konashi. Lattice parameters of $(U, Pu, Am, Np)O_{2-x}$, *Recent Advances in Actinide Science*, edited by Iain May, Rebecca Alvares and Nicholas Bryan. RSC Publishing, c2006, in *Actinides 2005*. 2005. Manchester.

16. Kato, M., et al., Physical properties and irradiation behavior analysis of Np- and Am-bearing MOX fuels. *Journal of Nuclear Science and Technology*, 2011. **48**: pp. 646–653.

17. Kato, M., et al., The effect of oxygen-to-metal ratio on melting temperature of uranium and plutonium mixed oxide fuel for fast reactor. *Transactions of the Atomic Energy Society of Japan*, 2008. **7**(4): pp. 420–428.

18. Komeno, A., et al., Oxygen potentials of PuO_{2-x}. *MRS Proceedings*, 2012. **1444**: pp.58–89.

19. Kato, M., et al., Oxygen potentials, oxygen diffusion coefficients and defect equilibria of nonstoichiometric $(U, Pu)O_{2\pm x}$. *Journal of Nuclear Materials*, 2017. **487**: pp. 424–432.

20. Kato, M., T. Uchida, and T. Sunaoshi, Measurement of oxygen chemical diffusion in PuO_{2-x} and analysis of oxygen diffusion in PuO_{2-x} and $(Pu, U)O_{2-x}$. *Physica Status Solidi (c)*, 2013. **10**(2): pp. 189–192.

21. Kato, M., et al., Thermal expansion measurement and heat capacity evaluation of hypo-stoichiometric $PuO_{2.00}$. *Journal of Nuclear Materials*, 2014. **451**(1–3): pp. 78–81.

22. Kato, M., et al., Oxygen potentials of plutonium and uranium mixed oxide. *Journal of Nuclear Materials*, 2005. **344**(1–3): pp. 235–239.

23. Kato, M., T. Tamura, and K. Konashi, Oxygen potentials of mixed oxide fuels for fast reactors. *Journal of Nuclear Materials*, 2009. **385**(2): pp. 419–423.

24. Kato, M., et al., Oxygen potential of $(U_{0.88}Pu_{0.12})O_{2\pm x}$ and $(U_{0.7}Pu_{0.3})O_{2\pm x}$ at high temperatures of 1673–1873K. *Journal of Nuclear Materials*, 2011. **414**(2): pp. 120–125.

25. Kato, M., et al., Measurement of oxygen potential of $(U_{0.8}Pu_{0.2})O_{2\pm x}$ at 1773 and 1873 K, and its analysis based on point defect chemistry. *Calphad*, 2011. **35**(4): pp. 623–626.

26. Kato, M., et al., Oxygen chemical diffusion in hypo-stoichiometric MOX. *Journal of Nuclear Materials*, 2009. **389**(3): pp. 416–419.

27. Kato, M., et al., Solidus and liquidus temperatures in the UO_2–PuO_2 system. *Journal of Nuclear Materials*, 2008. **373**(1–3): pp. 237–245.
28. Kato, M., et al., Solidus and liquidus of plutonium and uranium mixed oxide. *Journal of Alloys and Compounds*, 2008. **452**(1): pp. 48–53.
29. Kato, M., K. Konashi, and N. Nakae, Analysis of oxygen potential of $(U_{0.7}Pu_{0.3})O_{2\pm x}$ and $(U_{0.8}Pu_{0.2})O_{2\pm x}$ based on point defect chemistry. *Journal of Nuclear Materials*, 2009. **389**(1): pp. 164–169.
30. Kato, M. and K. Konashi, Lattice parameters of $(U, Pu, Am, Np)O_{2-x}$. *Journal of Nuclear Materials*, 2009. **385**(1): pp. 117–121.
31. Kato, M., et al., Self-radiation damage in plutonium and uranium mixed dioxide. *Journal of Nuclear Materials*, 2009. **393**(1): pp. 134–140.
32. Kato, M., et al., Thermal expansion measurement of $(U, Pu)O_{2-x}$ in oxygen partial pressure-controlled atmosphere. *Journal of Nuclear Materials*, 2016. **469**: pp. 223–227.
33. Kato, M. Melting temperatures of oxide fuel for fast reactors, in *International Congress on Advances in Nuclear Power Plants 2009*. 2009. Curran Associates, Inc, Atomic Energy Society of Japan (AESJ), Tokyo.
34. Morimoto, K., et al. The influence of Pu-content on thermal conductivities of (U, Pu) O_2 solid solution, in *Proceedings of International Conference on Fast Reactors and Related Fuel Cycles 2009*. 2009. Proceedings Series - International Atomic Energy Agency, Kyoto.
35. Hirooka, S. and M. Kato, Sound speeds in and mechanical properties of $(U, Pu)O_{2-x}$. *Journal of Nuclear Science and Technology*, 2017. **55**(3): pp. 356–362.
36. Hirooka, S., et al., Relative oxygen potential measurements of $(U, Pu)O_2$ with $Pu = 0.45$ and 0.68 and related defect formation energy. *Journal of Nuclear Materials*, 2022. **558**.
37. Kato, M., Oxygen potentials and defect chemistry in nonstoichiometric $(U, Pu) O_2$, in *Stoichiometry and Materials Science - When Numbers Matter*, A. Innocenti and N. Kamarulzaman, Eds. 2012. Intech, London, UK
38. Hirooka, S., et al., Oxygen potential measurement of $(U, Pu, Am)O_{2\pm x}$ and $(U, Pu, Am, Np)O_{2\pm x}$. *Journal of Nuclear Materials*, 2020. **542**: p. 152424.
39. Koizumi, M., et al., Development of a process for co-conversion of Pu-U nitrate mixed solutions to mixed oxide powder using microwave heating method. *Journal of Nuclear Science and Technology*, 1983. **20**(7): pp. 529–536.
40. Nakamichi, S., et al., Effect of O/M ratio on sintering behavior of $(Pu_{0.3}U_{0.7})O_{2-x}$. *Journal of Nuclear Materials*, 2020. **535**: p. 152188.
41. Takeuchi, K., M. Kato, and T. Sunaoshi, Influence of O/M ratio on sintering behavior of $(U_{0.8}, Pu_{0.2})O_{2\pm x}$. *Journal of Nuclear Materials*, 2011. **414**(2): pp. 156–160.

3 Properties

Masato Kato

CONTENTS

3.1 LATTICE PARAMETERS

U and Pu MOX, $(U, Pu)O_2$, has a fluorite-type crystal structure, which is shown in Figure 3.1. Metallic atoms are an array of face-centered cubic (f.c.c), and atoms of Pu and U form a substitutional solid solution. An oxygen atom is placed at the center of the tetrahedral site and constitutes a simple cubic sublattice. The filling rate in the fluorite crystal structure is low, which is about 63%, as compared with the 74% filling rate in the f.c.c. crystal. This suggests that the fluorite crystal can have many interstitial atoms and oxygen vacancy, and such defects significantly affect its lattice

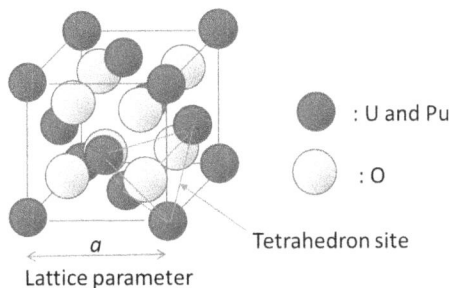

: U and Pu

: O

Tetrahedron site

a

Lattice parameter

FIGURE 3.1 Fluorite-type crystal structure of $(U, Pu)O_2$.

DOI: 10.1201/9781003298205-3

parameters. Therefore, lattice parameters are important basic units to express crystal structure and properties [1–14].

Metal atoms have a valence of +4 in stoichiometric compositions, assuming an ionic crystal. Lattice parameters decrease with increasing content of Pu, Am, and Np, which have smaller ionic radius than U^{4+}. In the nonstoichiometric region, the electrical neutrality condition and balance between the average valences of anions and cations are considered. A decrease in the O/M ratio increases the lattice parameters due to an increase in M^{3+} content, which has a larger ionic radius as shown in Figure 3.2.

In previous works, the lattice parameters of $(U, Pu, Am, Np)O_{2-x}$ were investigated, and the relational equation was derived as shown in Eq. (3.1). Theoretical density ρ_{th} can be obtained from Eq. (3.2) using a. It is also an important value that is used as the basis for heat generation density and temperature profile measurements in fuel performance analysis. Additional measurement data were added to the database, and the latest data set was made [2,3,4].

$$= 4/\sqrt{3} \cdot \{r_c (1 + 0.112x) + r_a\} \ \left(\overset{\circ}{A}\right) \tag{3.1}$$

$$\rho_{th} = \frac{4\bar{M}}{A_v \cdot a^3} \tag{3.2}$$

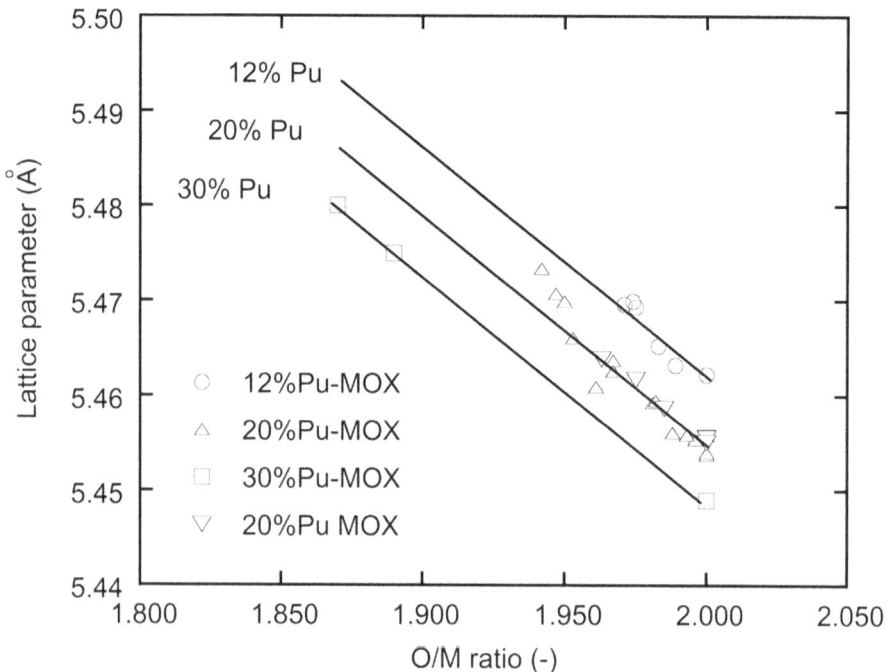

FIGURE 3.2 O/M dependence on the lattice parameter of $(U, P)O_2$.

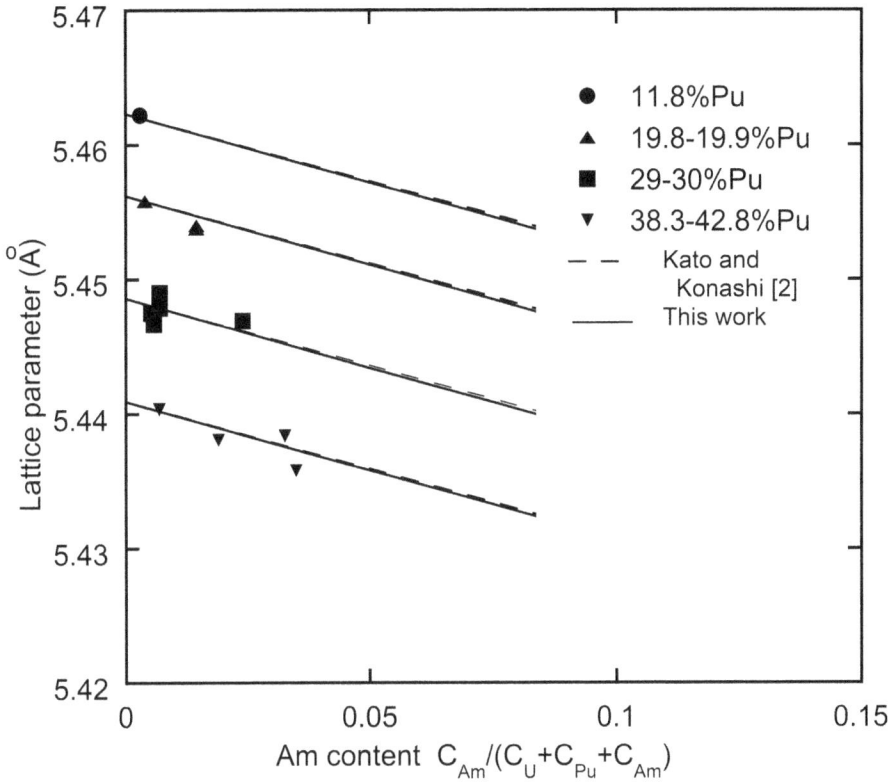

FIGURE 3.3   Am content depending on the lattice parameter of (U, Pu, Am, Np)O$_{2-x}$ (Kato and Konashi [2]).

The lattice parameters of (U, Pu, Am, Np)O$_{2-x}$ are described by Eq. (3.1) using ionic radius. The valence of actinide cation is assumed to be +4 [2]. Ionic radius affects the contribution of phonon conduction to thermal conductivity. Therefore, the lattice parameter-representation model is improved. It is reported that the oxygen potential of AmO$_2$ is higher than that of other actinide oxides [15]. It is assumed that the valence of Am is +3 when the O/M ratio is 2.00 in (U, Pu, Am, Np)O$_2$, and electroneutrality is maintained with the amount of U^{5+} similar to that of Am. The ionic radius of cation is described by the following equation:

$$r_c = r_{U4+} \cdot C_{U4+} + r_{U5+} \cdot C_{U5+} + r_{Pu} \cdot C_{Pu} + r_{Am3+} \cdot C_{Am} + r_{Np} \cdot C_{Np}, \text{ Å} \qquad (3.3)$$

Figure 3.3 shows the Am content dependence on the lattice parameters, and the ionic radius that is summarized in Table 3.1 was used. In this study, the values calculated using the improved model are presented with a solid line, whereas the experimental data that correspond to the improved model are presented with a broken line. The ionic radii summarized in Table 3.1 were used in thermal conductivity representation (see Section 3.9).

TABLE 3.1
Ionic Radius to Calculate a

Ionic Species	Ionic Radius Å
r_a	1.372
r_{U4+}	0.9972
r_{U5+}	0.9000
r_{Pu}	0.9642
r_{Am3+}	1.050
r_{Np}	0.9805

3.2 THERMAL EXPANSION

Dilatometers and high-temperature X-ray diffractometers were used in the thermal expansion measurements for $(U, Pu)O_{2-x}$ [10,16–55]. t has been reported that the data obtained by these two instruments differ due to the formation of lattice defects on other materials at high temperatures. However, no clear difference has been reported in the lattice expansion measurements of actinide oxides. In the measurement of the thermal expansion coefficient of MOXs, changes in the O/M ratio were observed before and after the measurement, and it was difficult to adjust the O/M ratio during the measurement. Thus, there were almost no data obtained for the O/M ratio. We measured the thermal expansion of $(U, Pu)O_{2-x}$ using a dilatometer in a P_{O_2}-controlled atmosphere to adjust the O/M ratio during measurement [6–9]. Figure 3.4 shows the measured data depending on the O/M ratio, and the data slightly increased with decreasing O/M ratio. The data were analyzed, and relational equations, Eqs. (3.4) and (3.5), were derived to describe a linear thermal expansion (LTE) as a function of Pu content, O/M ratio, and temperature [9] (Table 3.2).

$$LTE = \frac{\Delta L}{L_0} = a_0 + a_1 \cdot T + a_2 \cdot T^2 + a_3 \cdot T^3 \tag{3.4}$$

$$a_i = b_0 + b_1 \cdot C_{Pu} + b_2 \cdot x + b_3 \cdot C_{Pu}^2 + b_4 \cdot x^2 + b_5 \cdot C_{Pu} \cdot x \tag{3.5}$$

Minor actinide (MA)-bearing MOXs have been developed as fast reactor fuels. In development of nuclear fuels, thermal expansion needs to be determined to evaluate fuel performance, such as fuel and cladding mechanical interaction. However, the data of new nuclear fuels, such as MA-bearing MOXs, are limited. Therefore, the LTE values of MA-bearing MOXs were evaluated from the data of each stoichiometric actinide oxide, which are compared in Figure 3.5 [21,28,30,31]. The thermal expansion of AmO_2 was slightly lower compared with that of other oxides. However, there was no significant difference in thermal expansion values of MOXs based on the comparison.

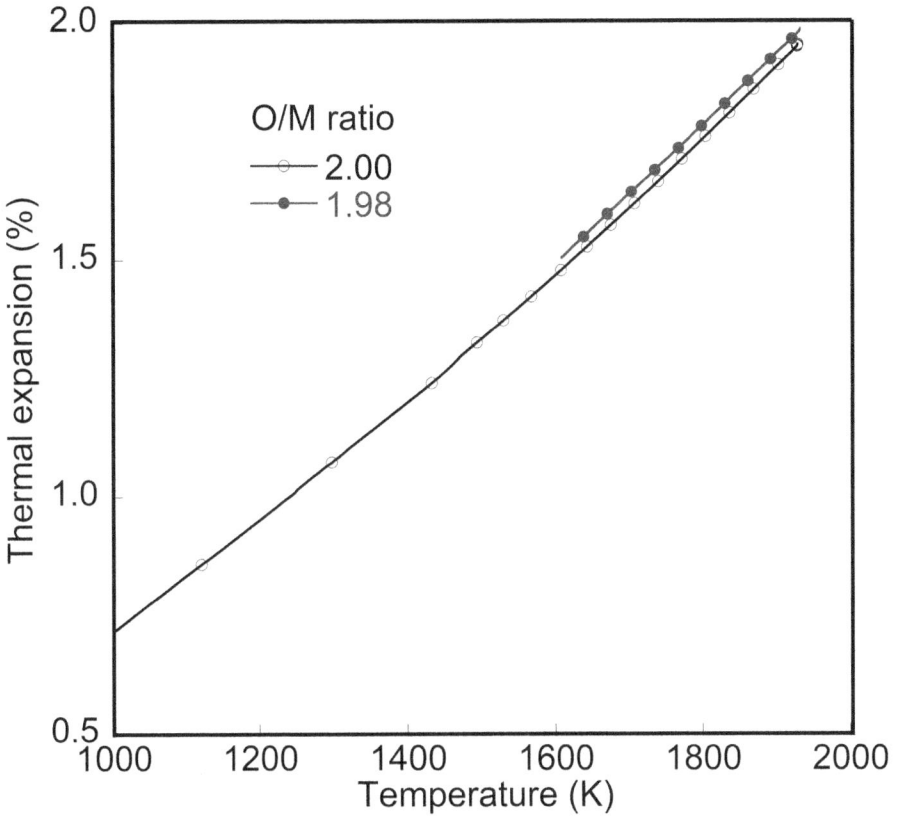

FIGURE 3.4 O/M dependence of $(U_{0.7}Pu_{0.3})O_{2-x}$ thermal expansion (Kato et al. [9]).

TABLE 3.2
Coefficients b_0, b_1, b_2, b_3, b_4, and b_5 for Eqs. (3.4) and (3.5)

a_n	b_0	b_1	b_2	b_3	b_4	b_5
a_0 ($\times 10^{-3}$)	−2.8809	0.0301	−4.3954	0.0156	−15.1759	2.5642
a_1 ($\times 10^{-6}$)	9.5024	−0.1864	15.8173	−0.0229	7.6258	−7.5789
a_2 ($\times 10^{-10}$)	2.0894	2.9483	−19.9227	−1.0355	73.8931	11.6442
a_3 ($\times 10^{-13}$)	4.4096	−1.4263	23.5638	0.0251	−54.751	−14.418

The thermal expansion of MA-bearing MOXs was evaluated using Eq. (3.6), and the parameters, a_i, are summarized in Table 3.3.

$$LTE = \left(C_U \cdot (LTE)_{UO_2} + C_{Pu} \cdot (LTE)_{PuO_2} + C_{Am} \cdot (LTE)_{AmO_2} + C_{Np} \cdot (LTE)_{NpO_2} \right)$$

$$\times (1 + 0.59614x) \tag{3.6}$$

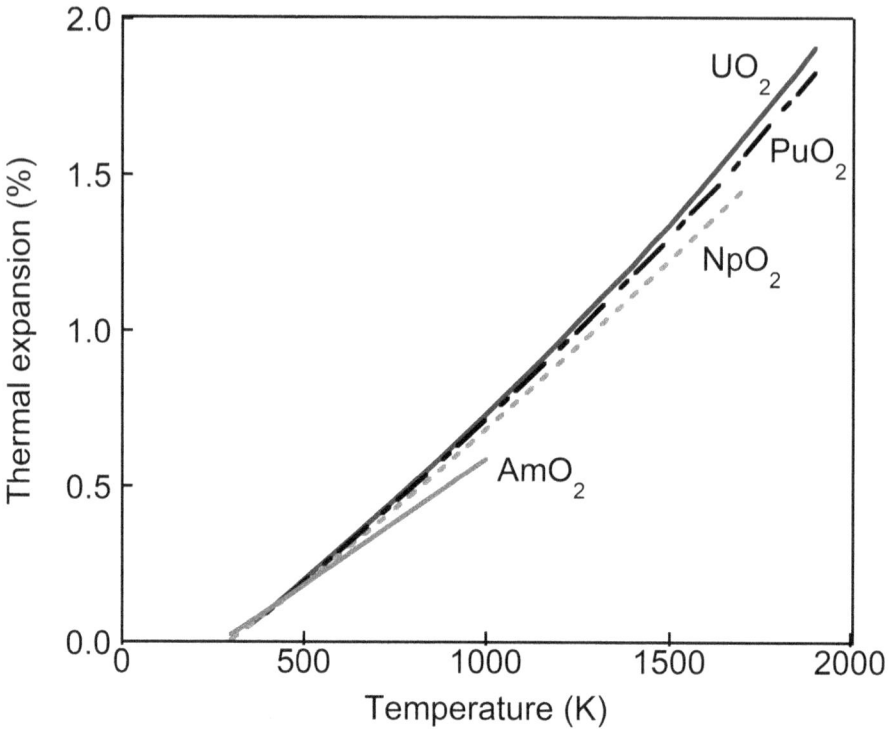

FIGURE 3.5 Comparison of the thermal expansion of actinide oxides (Gibby [21], Yamashita et al. [28], Minato et al. [30], Vauchy, Joly and Valot [31]).

TABLE 3.3
Coefficients a_0, a_1, a_2, and a_3, for Eq. (3.6)

	UO_2	PuO_2	AmO_2	NpO_2
a_0	−0.002881	−0.002835	−0.002425	−0.002880
a_1	9.5024 E-06	9.293E-06	8.063E-06	9.445E-06
a_2	2.089 E-10	4.002E-10	2.626E-11	4.7620E-10
a_3	4.4096 E-13	3.008E-13	-	-

The coefficient of thermal expansion α is also related to thermal expansion term in heat capacity. In this work, Eq. (3.6) was used to evaluate heat capacity, which is discussed in Section 3.8.

3.3   OXYGEN POTENTIAL

Oxygen potential is one of important thermodynamic data to evaluate chemical stability of oxide fuels. The oxygen potential of nonstoichiometric oxides can be determined as a function of the O/M ratio and temperature. Since the standard state is 1 atm, oxygen chemical potential in gas can be described as follows:

$$\mu(T, P_{O_2}) = \mu^{\circ}(T, P_{O_2}) + RT \ln P_{O_2} \tag{3.7}$$

On the other hand, oxygen chemical potential in oxides can be described as follows:

$$\mu(T, P_{O_2}) = \mu^{\circ}(T, P_{O_2}) + \Delta \bar{G}_{O_2} = \mu^{\circ}(T, P_{O_2}) + RT \ln a_{O_2} \tag{3.8}$$

Considering the equilibrium between atmospheric gas and oxides, Eqs. (3.7) and (3.8) result in the following equation:

$$\Delta \bar{G}_{O_2} = \mu(T, P_{O_2}) - \mu^{\circ}(T, P_{O_2}) = RT \ln P_{O_2} \tag{3.9}$$

Generally, the oxygen molar Gibbs free energy $\Delta \bar{G}_{O_2}$ is called the oxygen potential.

The ratio of O/M in the nonstoichiometric range shows a continuous variation depending on $\Delta \bar{G}_{O_2}$. Many studies on the oxygen potential of oxide fuels have been conducted using various methods [56–67], but the data have a high uncertainty of over 200 kJ/mol. Thus, it is difficult to evaluate the relationship among the O/M ratio, Pu content, T, and $\Delta \bar{G}_{O_2}$. We measured the oxygen potential of UO_2, PuO_2, (U, Pu) O_2, (U, Pu, Am)O_2, and (U, Pu, Am, Np)O_2 at high temperatures up to 1,973 K using a gas equilibrium method, through which more than 900 data points were obtained [69–81]. The data at 1,773 K depending on Pu content are plotted in Figure 3.6.

As shown in Figure 3.6, with increasing Pu content, the oxygen potential increased and expanded into the stoichiometric region. The data are plotted in double logarithmic of P_{O_2} versus x as shown in Figure 3.7.

In the plot shown in Figure 3.7, the relationship of $x \propto P_{O_2}^{1/n}$ is observed, where n is the characteristic number depending on the type of lattice defect. It can be observed from Figure 3.7 that n changed from -2 to $+2$ in hypo- and hyperstoichiometric compositions. In the reduction region, it was observed that n equals -4 in MOXs containing 20%, 70%, and 100% Pu, and n equals -3 in MOXs containing 30% and 45% Pu. The relationship between n and Pu content was evaluated using all experimental data, and $\Delta \bar{G}_{O_2}$ was determined as a function of Pu content, MA content, and O/M ratio. These evaluated results were used in the analysis based on defect chemistry in this work.

The physical properties of nonstoichiometric compounds, such as oxygen potential, electrical conductivity, and oxygen diffusion, have been studied based on defect chemistry [73,82–85]. Evaluations of oxygen diffusion and electrical conductivity are discussed in Sections 3.4 and 3.6.

Recently, $\Delta \bar{G}_{O_2}$ of MOXs containing high Pu content (about 50% and 70% Pu) was measured [79]. In addition, the effect of Am and Np content on oxygen potential data was quantitatively investigated [81]. These data were added to the database. The oxygen potential data of (U, Pu)O_2 was analyzed using the latest data set to construct a Brouwer diagram.

Two kinds of Brouwer diagram are proposed with a predominant intrinsic defect and Frenkel defect in near stoichiometric composition. The Brouwer diagrams of PuO_2 and CeO_2 are shown in Figure 3.8, which were reproduced from a previous work [86].

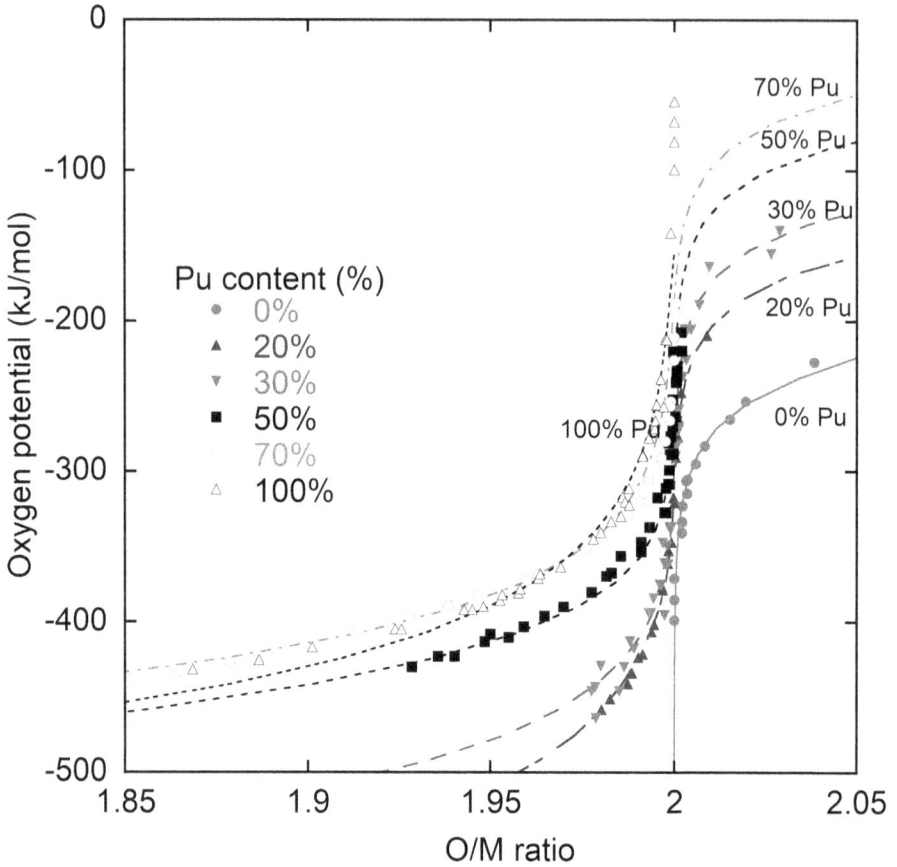

FIGURE 3.6   Oxygen potential of (U, Pu)O$_2$ at 1,773 K (Watanabe, Kato and Sunaoshi [69], Komeno et al. [72], Kato et al. [76], Kato et al. [77], Hirooka et al. [79]).

It was assumed that the defect concentration of electron hole and Frenkel defect was dominant in PuO$_2$ and CeO$_2$, respectively. In these materials, electronic conduction and ionic conduction are the expected conduction mechanisms near stoichiometric composition. Analysis results consistently explained about oxygen potential, oxygen diffusion coefficients, defect equilibria, and thermal properties in the oxides. Electrical conduction measurement results revealed that intrinsic defect was also dominant in UO$_2$ and (U, Pu)O$_2$. Hence, the Brouwer diagram of (U, Pu)O$_2$ was evaluated similarly as that of PuO$_2$.

In the near stoichiometric region, defect equilibria in the following reactions were considered [73]:

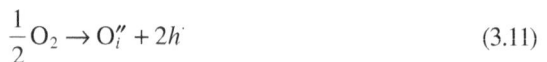

$$O_O^\times \rightarrow V_o'' + 2e' + \frac{1}{2}O_2 \tag{3.10}$$

$$\frac{1}{2}O_2 \rightarrow O_i'' + 2h^{\cdot} \tag{3.11}$$

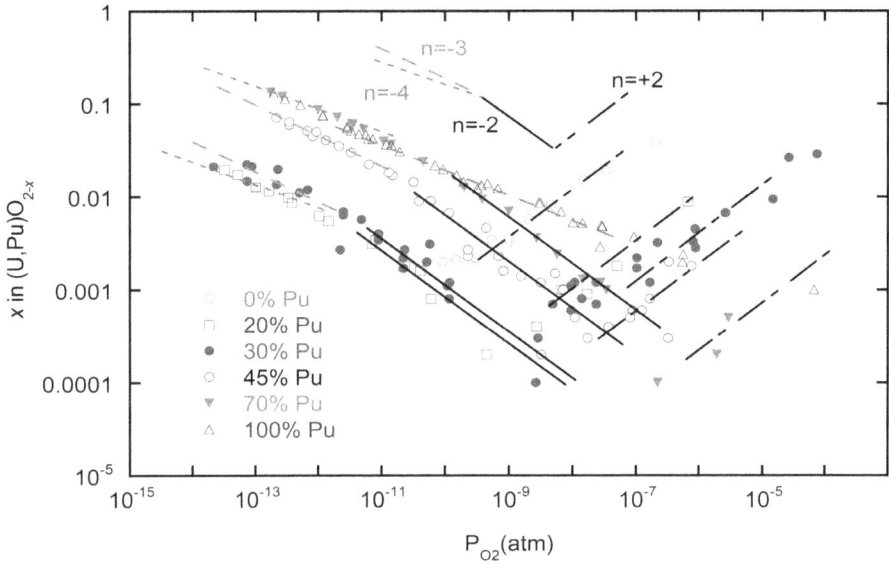

FIGURE 3.7 Oxygen partial pressure dependence on the deviation x in $(U, Pu)O_{2 \pm x}$ at 1,773 K (Watanabe, Kato and Sunaoshi [69], Komeno et al. [72], Kato et al. [76], Kato et al. [77], Hirooka et al. [79]).

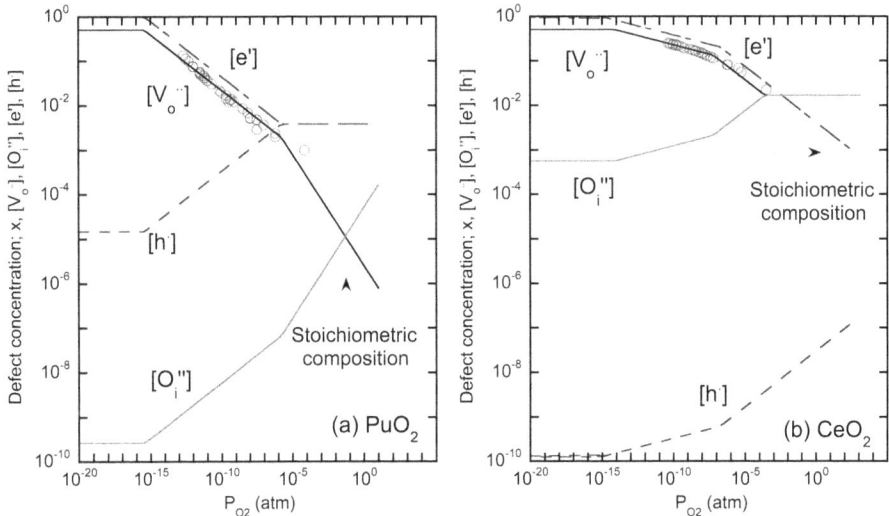

FIGURE 3.8 Brouwer diagrams of (a) PuO_2 and (b) CeO_2 at 1,773 K.

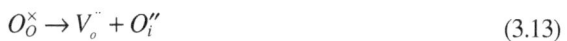

$$null \rightarrow e' + h^{\cdot} \qquad (3.12)$$

$$O_O^x \rightarrow V_o^{\cdot\cdot} + O_i'' \qquad (3.13)$$

The equilibrium constants in the above defect reactions can be described using the following equations:

$$K_V = \left[V_o^{\cdot\cdot} \right]\left[e' \right]^2 P_{O_2}^{-1/2}$$

(3.14)

$$K_O = \left[O_i'' \right]\left[h^{\cdot} \right]^2 P_{O_2}^{-1/2}$$

(3.15)

$$K_i = \left[e' \right]\left[h^{\cdot} \right]$$

(3.16)

$$K_F = \left[V_o^{\cdot\cdot} \right]\left[O_i'' \right]$$

(3.17)

Previous studies on electrical measurements showed that electronic conduction is dominant near stoichiometric $(U, Pu)O_2$, which means that the defect concentration of e' and $h^{\cdot}$ dominate that of $V_o^{\cdot\cdot}$ and O_i''. Therefore, $\left[e' \right] = \left[h^{\cdot} \right]$ was assumed near the stoichiometric region. The following equations were obtained from Eqs. (3.14) to (3.17).

$$\left[e' \right] = \left[h^{\cdot} \right] = K_i^{1/2}$$

(3.18)

$$\left[O_i'' \right] = \left(K_O / K_i \right) \cdot P_{O_2}^{1/2}$$

(3.19)

$$\left[V_o^{\cdot\cdot} \right] = \left(K_V / K_i \right) \cdot P_{O_2}^{-1/2}$$

(3.20)

$$K_F = K_V \cdot K_O / K_i^2$$

(3.21)

In the reducing region, the oxygen potential experimental data showed that deviation, x, is proportional to $P_{O_2}^{-1/4}$, $P_{O_2}^{-1/3}$, and $P_{O_2}^{-1/4}$ in the regions of 0%–20%, 30%–70%, and 100% Pu content, respectively. The defect reactions presented in Eqs. (3.23) and (3.24) were assumed in the regions of $x \propto P_{O_2}^{-1/4}$ and $P_{O_2}^{-1/3}$, respectively.

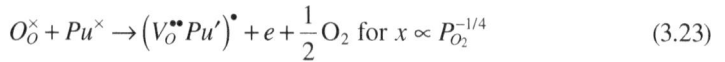

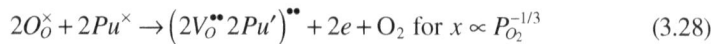

$$O_O^{\times} + Pu^{\times} \rightarrow \left(V_O^{\cdot\cdot} Pu' \right)^{\cdot} + e + \frac{1}{2} O_2 \text{ for } x \propto P_{O_2}^{-1/4}$$

(3.23)

$$2O_O^{\times} + 2Pu^{\times} \rightarrow \left(2V_O^{\cdot\cdot} 2Pu' \right)^{\cdot\cdot} + 2e + O_2 \text{ for } x \propto P_{O_2}^{-1/3}$$

(3.28)

The $\left[V_o^{\cdot\cdot} \right]$ can be represented by the following equations:

$$\left[V^{\cdot\cdot} \right] = \left(K_{-1/4} \right)^{1/2} P_{O_2}^{-1/4} \qquad \text{for } n = -4,$$

(3.29)

$$\left[V^{\cdot\cdot} \right] = \left(2K_{-1/3} \right)^{1/3} P_{O_2}^{-1/3} \qquad \text{for } n = -3,$$

(3.30)

In the reducing region, $[e']$ was equivalent to $2[Vo^{..}]$. At the boundary between the reducing and near stoichiometric regions, the following equations are obtained:

$$K_i^{1/2} = 2K_{-1/4}P_{O_2}^{-1/4} \tag{3.31}$$

$$K_i^{1/2} = 2(2K_{-1/3})^{1/3} P_{O_2}^{-1/3} \tag{3.32}$$

The relationship $x \propto P_{O_2}^{-1/4}$ was observed in the U- and Pu-rich composition region, and equilibrium constants are defined as $\left(K_{-\frac{1}{4}}\right)_U$ and $\left(K_{-\frac{1}{4}}\right)_{Pu}$, respectively. In the region of $x \propto P_{O_2}^{-1/4}$, $[V_o^{..}]$ is equal to $2[e']$, and Eq. (3.33) is obtained.

$$[e']_{-1/4} = 2(K_{-1/4})^{1/2} P_{O_2}^{-1/4} \tag{3.33}$$

Here, it was assumed that $[e']_{-1/4}$ is equal to $[e']_{Sto.}$ at P_{O_2} at the boundary between the reducing and near stoichiometric regions. The following equations are obtained:

$$\Delta H_{-1/4} = 1/2\Delta H_i + \Delta H_{n=-2} \tag{3.34}$$

$$\Delta S_{n=-4} = 1/2\Delta S_i + \Delta S_{-1/2} - 1/2R\Delta \ln(4) \tag{3.35}$$

In the case of Eq. (3.32), the following equations are derived:

$$\Delta H_{-1/3} = \Delta H_i/2 - 2\Delta H_{-1/2} \tag{3.36}$$

$$\Delta S_{n-1/3} = \Delta S_i/2 + 2\Delta S_{-1/2} - 1/2 \cdot R \cdot \ln(16) \tag{3.37}$$

The equilibrium constant is written as follows:

$$K_m = \exp(\Delta S_m/R) \cdot \exp(-\Delta H_m/RT) \tag{3.22}$$

Deviation, x, is recognized to be equivalent to $[V_o^{..}]$ or $[O_i'']$. The oxygen potential experimental data were fitted by Eqs. (3.3)–(3.19), (3.20), (3.29), and (3.30), and then ΔS_m and ΔH_m were determined. ΔS_i and ΔH_i for UO$_2$, (U, Pu)O$_2$, and PuO$_2$ were analytically determined as follows:

UO$_2$: $\Delta S_i = $ 85 J/mol K and $\Delta H_i = $ 300 kJ/mol K

MOX : $\Delta S_i = \left(71.1 + 67.35 \cdot C_{Pu} + 108.0 \cdot C_{Pu}^2 - 198.7 \cdot C_{Pu}^3\right)$ J/mol K and

$\Delta H_i = $ 278 kJ/mol K

PuO$_2$: $\Delta S_i = $ 105 J/mol K and $\Delta H_i = $ 350 kJ/mol K

Assuming that defect equilibria are maintained in all composition range of the non-stoichiometric oxide, a Brouwer diagram can be constructed. Brouwer diagrams of $(U_{0.88}Pu_{0.12})O_2$, $(U_{0.8}Pu_{0.2})O_2$, and $(U_{0.7}Pu_{0.3})O_2$ were obtained as shown in Figure 3.9.

Finally, the following equation was derived to represent the O/M ratio in (U, Pu) O_{2-x} as a function of Pu content, oxygen partial pressure, and temperature. The calculation results are shown with lines in Figure 3.6 and consistent with the experimental data.

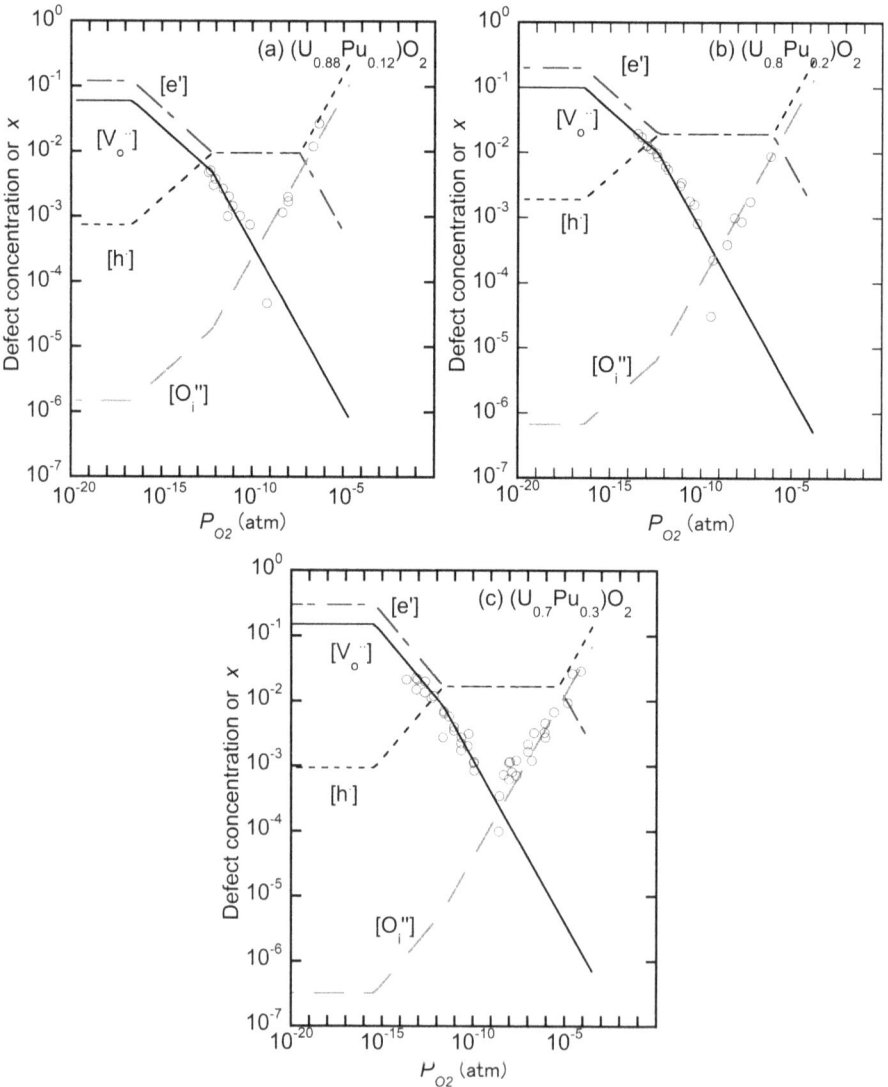

FIGURE 3.9 Brouwer diagrams of (a) $(U_{0.88}Pu_{0.12})O_2$, (b) $(U_{0.8}Pu_{0.2})O_2$, and (c) $(U_{0.7}Pu_{0.3})O_2$.

$$O/M = 2 + [Oi''] - [Vo\ddot{}]$$

$$= 2 + \left[\left\{ (K_{1/2})_U \cdot P_{O_2}^{\frac{1}{2}} \right\}^{-5} + \left\{ (K_{1/2})_{Pu} \cdot P_{O_2}^{\frac{1}{2}} \right\}^{-5} \right]^{-1/5} - \left[\left\{ (K_{-1/2})_U \cdot P_{O_2}^{-\frac{1}{2}} \right\}^{-5} \right.$$

$$+ \left\{ (K_{-1/2})_M \cdot P_{O_2}^{-\frac{1}{2}} \right\}^{-5} + \left\{ (K_{-1/2})_{Pu} \cdot P_{O_2}^{-\frac{1}{2}} \right\}^{-5} + \left\{ ((K_{-1/4})_U)^{1/2} \cdot P_{O_2}^{-\frac{1}{4}} \right\}^{-5}$$

$$+ \left\{ (2 \cdot K_{-1/3})^{1/3} \cdot P_{O_2}^{-\frac{1}{3}} \right\}^{-5} + \left\{ ((K_{-1/4})_{Pu})^{1/2} \cdot P_{O_2}^{-\frac{1}{4}} \right\}^{-5} + \left\{ C_{Pu}'/2 \right\}^{-5} \right]^{-1/5}$$

$$= 2 + \left[\left[\left\{ exp\left(\frac{\left(-20.0 - 112.2 \cdot C_{Pu}' + 58.2 \cdot C_{Pu}'^2 \right)}{R} \right) \cdot exp\left(\frac{105,000}{RT} \right) \cdot P_{O_2}^{\frac{1}{2}} \right\}^{-5} \right. \right.$$

$$+ \left\{ exp\left(\frac{\left(2595.3 - 2,600 \cdot C_{Pu}' \right)}{R} \right) \cdot exp\left(\frac{-159,300}{RT} \right) \cdot P_{O_2}^{\frac{1}{2}} \right\}^{-5} \right]^{-1/5}$$

$$- \left[\left\{ exp\left(\frac{\left(42.3 + 48.3 \cdot C_{Pu}' + 110.0 \cdot C_{Pu}^2 + 99.4 \cdot C_{Pu}'^3 \right)}{R} \right) \cdot exp\left(\frac{-372,000}{RT} \right) \cdot P_{O_2}^{-\frac{1}{2}} \right\}^{-5} \right.$$

$$+ \left\{ exp\left(\frac{\left(125.8 - 745.1 \cdot C_{Pu}' + 2678.5 \cdot C_{Pu}'^2 - 3683.4 \cdot C_{Pu}'^3 + 1792.7 \cdot C_{Pu}^4 \right)}{R} \right) \right.$$

$$\left. \cdot exp\left(\frac{-376,500}{RT} \right) \cdot P_{O_2}^{-\frac{1}{2}} \right\}^{-5}$$

$$+ \left\{ exp\left(\frac{\left(91.6 - 63.7 \cdot C_{Pu} - 54.0 \cdot C_{Pu}^2 + 99.4 \cdot C_{Pu}'^3 \right)}{R} \right) \cdot exp\left(\frac{-306,000}{RT} \right) \cdot P_{O_2}^{-\frac{1}{2}} \right\}^{-5}$$

$$+ \left\{ \left(exp\left(\frac{\left(72.1 + 82.0 \cdot C_{Pu}' + 164.0 \cdot C_{Pu}'^2 \right)}{R} \right) \cdot exp\left(\frac{-511,000}{RT} \right) \right)^{1/2} \cdot P_{O_2}^{-\frac{1}{4}} \right\}^{-5}$$

$$+ \left\{ 2 \cdot exp\left(\frac{\left(275.7 - 1456.5 \cdot C_{Pu}' + 5411.0 \cdot C_{Pu}'^2 - 7466.2 \cdot C_{Pu}'^3 + 3585.3 \cdot C_{Pu}'^4 \right)}{R} \right) \right.$$

$$\tag{3.38}$$

$$\cdot exp\left(\frac{-892,000}{RT}\right)\Bigg)^{1/3}\cdot P_{O_2}^{-\frac{1}{3}}\Bigg\}^{-5}$$

$$+\left\{\left(exp\left(\frac{\left(121.4-30.0\cdot C_{Pu}{}'\right)}{R}\right)\cdot exp\left(\frac{-445,000}{RT}\right)\right)^{1/2}\cdot P_{O_2}^{-\frac{1}{4}}\right\}^{-5}+\left\{C_{Pu}{}'/2\right\}^{-5}\right]^{-1/5}$$

Equation (3.38) was rewritten as Eqs. (3.39–3.41) to calculate P_{O_2} in the region of O/M = 2.000.

O/M = 2.000

$$P_{O_2}=\Bigg[\Bigg[\left\{exp\left(\frac{\left(42.3+48.3\cdot C_{Pu}{}'+110.0\cdot C_{Pu}{}'^2+99.4\cdot C_{Pu}{}'^3\right)}{R}\right)\cdot exp\left(\frac{-372,000}{RT}\right)\right\}^{-5}$$

$$+\left\{exp\left(\frac{\left(125.8-745.1\cdot C_{Pu}{}'+2678.5\cdot C_{Pu}{}'^2-3683.4\cdot C_{Pu}{}'^3+1792.7\cdot C_{Pu}{}'^4\right)}{R}\right)\right.$$

$$\left.\cdot exp\left(\frac{-376,500}{RT}\right)\right\}^{-5}\Bigg\}$$

$$+\left\{exp\left(\frac{\left(91.6-63.7\cdot C_{Pu}{}'-54.0\cdot C_{Pu}{}'^2+99.4\cdot C_{Pu}{}'^3\right)}{R}\right)\cdot exp\left(\frac{-306,000}{RT}\right)\right\}^{-5}\Bigg]^{-\frac{1}{5}}$$

$$\Bigg[\Bigg[\left\{exp\left(\frac{\left(-20.0-112.2\cdot C_{Pu}{}'+58.2\cdot C_{Pu}{}'^2\right)}{R}\right)\cdot exp\left(\frac{105,000}{RT}\right)\right\}^{-5}$$

$$+\left\{exp\left(\frac{2595.3-2600\cdot C_{Pu}{}'}{R}\right)\cdot exp\left(\frac{-159,300}{RT}\right)\right\}^{-5}\Bigg]^{-1/5} \tag{3.39}$$

The effect of Am and Np addition on the oxygen potential was investigated by Hirooka et al. [81]. The effect of Am and Np is described as follows:

$$C_{Pu}{}'=C_{Pu}+2.5C_{Am}+0.5C_{Np} \tag{3.40}$$

TABLE 3.4
Comparison of Formation Energies

Type of Defect	UO$_2$	MOX	PuO$_2$
$e' + h\dot{}$	300 (3.1)	278 (2.9)	350 (3.6)
$Vo\ddot{} + Oi''$	421 (4.4)	317 (3.3)	465 (4.8)
$Vo\ddot{}$	481 (5.0)	377 (3.9)	306 (3.2)
Oi''	−60 (−0.6)	−105 (−1.1)	159.3 (1.7)

kJ/mol (eV).

Matsumoto et al. measured the oxygen potential of (Pu, Am)O$_{2-x}$ [71]. The data changed in two stages, which were caused by the effects of the Am and Pu separately. Equations (3.39) and (3.40), which were derived in this study, cannot yet represent the data of (Pu, Am)O$_{2-x}$, but the equations can describe the data of (U, Pu, Am, Np)O$_2$ with a standard deviation of 40 kJ/mol.

The defect formation energies obtained by this analysis are summarized in Table 3.4. The formation energies of $e' + h\dot{}$, $Vo\ddot{} + Oi''$, $Vo\ddot{}$, and Oi'' in MOX were evaluated to be 278 kJ/mol (2.9 eV), 317 kJ/mol (3.3 eV), 377 kJ/mol (3.9 eV), and −105 kJ/mol (−1.1 eV), respectively. The formation energies of $e' + h\dot{}$ and $Vo\ddot{} + Oi''$ in MOX were smaller than those of UO$_2$ and PuO$_2$. Oxygen Frenkel defect formation energies have been reported in UO$_2$ [87–105] and PuO$_2$ [97, 105–111], and their data are scattered. The average values are 410 and 450 kJ/mol, respectively, in UO$_2$ and PuO$_2$, which are consistent with the current data, 421 and 465 kJ/mol, respectively.

3.4 OXYGEN DIFFUSION COEFFICIENTS

Masato Kato and Masashi Watanabe

It is necessary to know the behavior of oxygen in MOXs to analyze fuel performance during MOX irradiation and to control the sintering behavior and O/M ratio in pellet production process. Therefore, studies have been conducted on the self-diffusion coefficient D^* and chemical diffusion coefficient $\tilde{D}$ of oxygen in MOXs. D^* is not accompanied by a change in oxygen content, whereas $\tilde{D}$ is accompanied by a change in oxygen content. Considering a fluorite-structured single-phase sample having an infinitely cylindrical geometry, D^* and $\tilde{D}$ can be obtained by solving the following equations [112]:

$$\frac{C - C_\mathrm{f}}{C_\mathrm{i} - C_\mathrm{f}} = 1 - \sum_{n=1}^{\infty} \frac{4\left(\dfrac{a}{D^*}\right)^2 \exp\left(-\beta_n^2 D^* t / a^2\right)}{\beta_n^2 \left(\beta_n^2 - \left(\dfrac{a}{D^*}\right)^2\right)} \tag{3.41}$$

$$\frac{C - C_f}{C_i - C_f} = \sum_{n=1}^{\infty} \frac{4 \exp\left(-\beta_n^2 \tilde{D} / a^2\right)}{\beta_n^2}, \tag{3.42}$$

Here, C_i, C_f, a, t, and the β_n are the initial concentration, final concentration, radius, time, and roots of the equation $J_o(x) = 0$, respectively, where $J_o(x)$ is the Bessel function of zero order. If oxygen content can be determined from the change in the weight of the sample, diffusion coefficient can be obtained using thermogravimetry.

$\tilde{D}$ was measured using thermogravimetry and dilatometry [113–146]. In early studies, chemical diffusion coefficient was determined using an oxidation curve from a low O/M ratio to O/M = 2.00. Kato et al. investigated the oxygen diffusion coefficient from oxidation and reduction curves [116]. Their results showed that the oxidation process did not follow the chemical diffusion mechanism, and $\tilde{D}$ was determined from reduction curves. Watanabe et al. measured oxygen chemical and self-diffusion coefficient in $(U, Pu)O_{2-x}$ using gravimetry [113,114]. The measurement of D^* in MOXs was conducted under Ar/H_2 atmosphere with added moisture that was made from water consisting of the isotope ^{18}O. The relationship between oxygen potential and oxygen diffusion coefficients was presented. The migration energy of oxygen vacancy and interstitial oxygen were 60 and 112.5 kJ/mol, respectively, and all parameters in Eqs. (3.43) and (3.44) were determined.

$$D^* = D_{V_o}^0 [V_o] \exp\left(-\frac{\Delta H_{V_o}^m}{RT}\right) + 2D_{O_i}^0 [O_i] \exp\left(-\frac{\Delta H_{O_i}^m}{RT}\right) \tag{3.43}$$

$$\tilde{D} = \frac{2 \pm x}{2x} D^* \left(\pm \frac{\partial \log P_{O_2}}{\partial \log x}\right), \tag{3.44}$$

The crosspoint between $\left[O_i''\right]$ and $\left[V_o''\right]$ in Figure 3.10a indicates stoichiometric composition. Oxygen diffusion coefficients are shown in Figure 3.10b, which were calculated using Eq. (3.38). The lowest oxygen self-diffusion coefficient value was obtained at P_{O_2}, which was different from the stoichiometric composition. The lowest value corresponds to O/M = 1.999. The deviation was due to the higher diffusion coefficient of interstitial oxygen compared with oxygen vacancy. Calculation results are consistent with the experimental data with a standard deviation of 0.6 in log D.

Vauchy et al. [145] measured oxygen self-diffusion coefficient in $(U_{0.55}Pu_{0.45})O_2$ using the isotope ^{18}O. The coefficient was determined to be $4.5 \times 10^{-15} - 3.7 \times 10^{-14}$ m²/s at 1,273 K in an atmosphere of $P_{O_2} = 3.5 \times 10^{-16} - 2.7 \times 10^{-18}$ atm. The O/M ratio of the sample was evaluated to be 2.00 ± 0.001. Based on the current model, a self-diffusion coefficient of $1.1 \times 10^{-14} - 8.1 \times 10^{-14}$ m²/s could be obtained at O/M = 1.9999–1.9992. The calculated values were consistent with the experimental values.

Abnormally rapid O/M change of MOX pellets with a low O/M ratio has been reported near room temperature [147–149]. Woodley and Gibby investigated the oxidation rate of MOXs with a low O/M ratio and reported their rapid oxidation [147] that was shown to be affected by moisture. Suzuki et al. investigated oxidation behavior at temperatures of 333–473 K, analyzed two-phase oxidation models with different O/M ratios, and evaluated the activation energy of chemical diffusion to 73.4 kJ/mol [148]. Storage at 323 K or less increases the O/M ratio to about 0.01 after 3 months. Tanaka et al. also examined oxidation at 673–873 K and reported that rapid oxidation occurred [149]. The oxidation rate of these low-O/M MOXs cannot

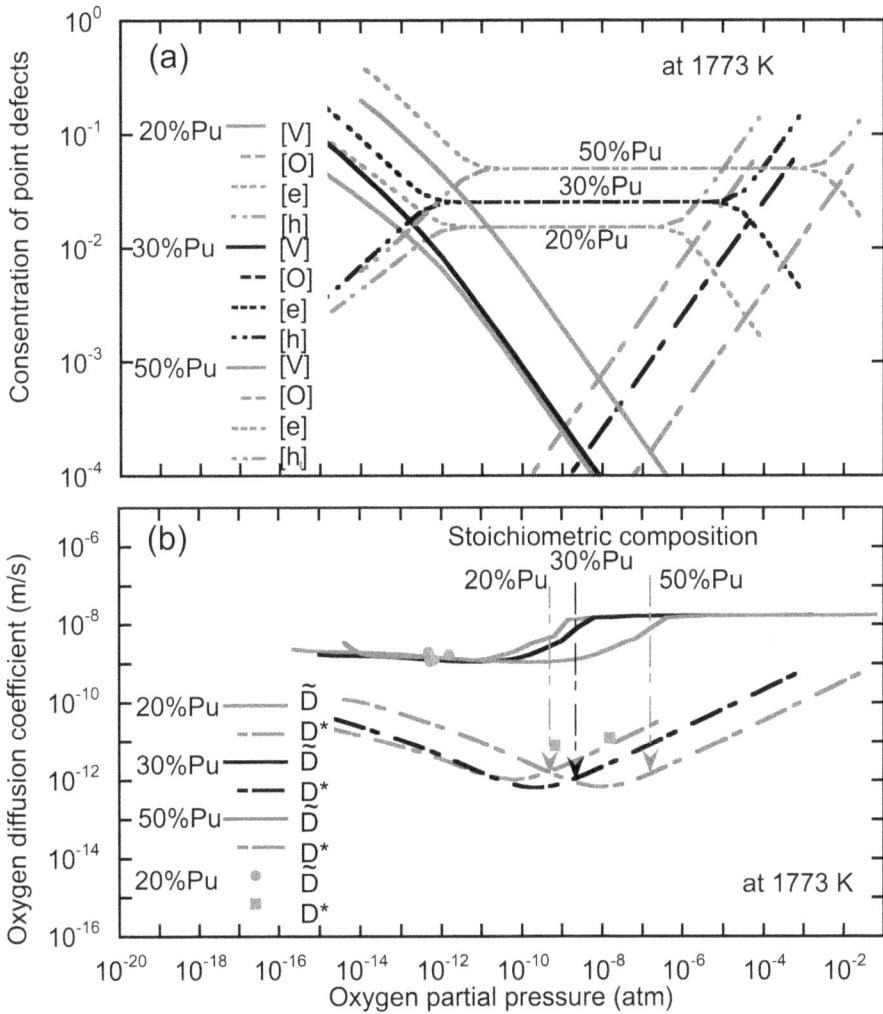

FIGURE 3.10 (a) Brouwer diagrams and (b) oxygen diffusion coefficients of $(U_{0.8}Pu_{0.2})O_2$, $(U_{0.7}Pu_{0.3})O_2$, and $(U_{0.5}Pu_{0.5})O_2$ at 1,773 K.

be explained by the diffusion coefficient shown in the equation. The mechanism of low-temperature oxidation behavior of low-O/M MOXs is still unknown.

3.5 MELTING TEMPERATURE

Early studies on melting temperature were conducted using the tungsten filament method. This measurement yields an incorrect melting temperature because of the vaporization of the specimen. Then the melting temperature of $(U, Pu)O_{2.00}$ was measured using a tungsten capsule, and the phase diagram in UO_2–PuO_2 solid solution was reported [150–169] as shown with a solid line in Figure 3.11.

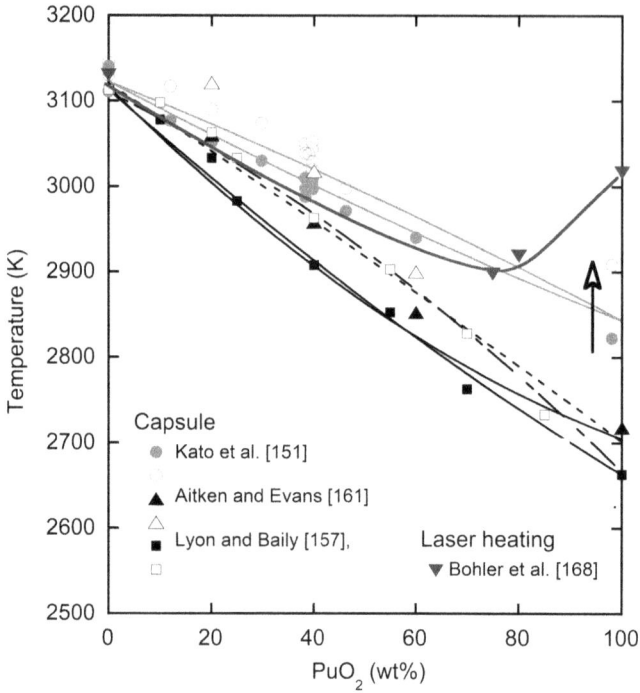

FIGURE 3.11   Melting temperature in UO_2–PuO_2 solid solution (closed symbol: Solidus, and open symbol: Liquidus) (Kato et al. [151], Aitken and Evans [161], Lyon and Baily [157], Böhler et al. [168]).

Kato et al. reported that the reaction between tungsten capsule and MOXs having a high Pu content (more than 30%) affected melting temperature measurements. The melting temperatures of MOXs with high Pu content, which were measured using the rhenium inner capsule [150,151], are shown in the current database. A rhenium inner was used to avoid the reaction between the MOX sample and capsule material, and it was shown that using the rhenium inner, the melting temperature of PuO_2 was 2,843 K, which was about 150 K higher than old data. However, the solidus temperature of PuO_2 was about 100 K different from the liquidus temperature. In addition, a reaction between PuO_2 and Re was still observed even using the Re inner in the melting measurement of PuO_2. Moreover, Manara et al. employed a laser-heating method to determine the melting point of PuO_2 and reported that the melting point of PuO_2 was 3,014 K [164,166]. Böhler et al. reported the revised phase diagram in the UO_2–PuO_2 system through melting temperature measurement of MOXs using the laser-heating method [167,168].

In a previous work, solidus and liquidus temperatures were represented by an ideal solid solution model using Eqs. (3.45)–(3.52). All parameters are summarized in Table 3.5.

$$\frac{\Delta h_m\left(UO_2\right)}{R}\left(\frac{1}{T_m\left(UO_2\right)}-\frac{1}{T_m}\right)=\ln\left(\frac{x^l\left(UO_2\right)}{x^s\left(UO_2\right)}\right) \qquad (3.45)$$

TABLE 3.5

Comparison of Melting Temperatures and Heat of Fusion of UO_2, PuO_2, AmO_2, NpO_2, and $PuO_{1.7}$

Melting Temperature (K)					Heat of Fusion (kJ/mol)				
UO_2	PuO_2	AmO_2	NpO_2	$PuO_{1.7}$	UO_2	PuO_2	AmO_2	NpO_2	$PuO_{1.7}$
3,128	2,839	2,813	2,820	3,051	101.4	92.0	91.2	70.3	479.3

$$\frac{\Delta h_m(PuO_2)}{R}\left(\frac{1}{T_m(PuO_2)} - \frac{1}{T_m}\right) = \ln\left(\frac{x^l(PuO_2)}{x^s(PuO_2)}\right) \tag{3.46}$$

$$\frac{\Delta h_m(AmO_2)}{R}\left(\frac{1}{T_m(AmO_2)} - \frac{1}{T_m}\right) = \ln\left(\frac{x^l(AmO_2)}{x^s(AmO_2)}\right) \tag{3.47}$$

$$\frac{\Delta h_m(PuO_{1.7})}{R}\left(\frac{1}{T_m(PuO_{1.7})} - \frac{1}{T_m}\right) = \ln\left(\frac{x^l(PuO_{1.7})}{x^s(PuO_{1.7})}\right), \tag{3.48}$$

$$x^l(UO_2) + x^l PuO_2 + x^l AmO_2 + x^l PuO_{1.7} = 1 \tag{3.49}$$

$$x^s(UO_2) + x^s(PuO_2) + x^s(AmO_2) + x^s(PuO_{1.7}) = 1 \tag{3.50}$$

$$x^l(UO_2) + x^l(PuO_2) + x^l(AmO_2) + x^l(PuO_{1.7}) = 1 \tag{3.51}$$

$$\Delta h_m(t) = 3RT_m, \tag{3.52}$$

Guéneau et al. assessed the phase diagram in the U–Pu–O system using a thermodynamic database [169] and made an excellent representation of the melting behavior.

Oxygen potential is one of the thermodynamic properties used for evaluating the chemical stability of oxides and is considered to be related to melting temperature. Oxygen partial pressure was calculated at solidus temperature T_m using Eq. (3.38). As shown in Figure 3.12, the $\ln P_{O_2}$ at melting temperature was plotted against $1/T_m$. The relationship can be represented using the following equations:

$$\ln P_{O_2} = 617,000/T_m - 209.5 \tag{3.53}$$

$$\ln P_{O_2} = 195,700/T_m - 68.0 \tag{3.54}$$

$$\ln P_{O_2} = -404,000/T_m + 138.4 \tag{3.55}$$

Equations (3.38) and (3.53)–(3.55) could describe solidus temperature depending on the composition. Solidus temperatures, depending on Pu content and O/M ratio system, are plotted in Figures 3.13 and 3.14. In addition, calculation results are shown in the figures with lines. The calculations represent the experimental data. The lowest

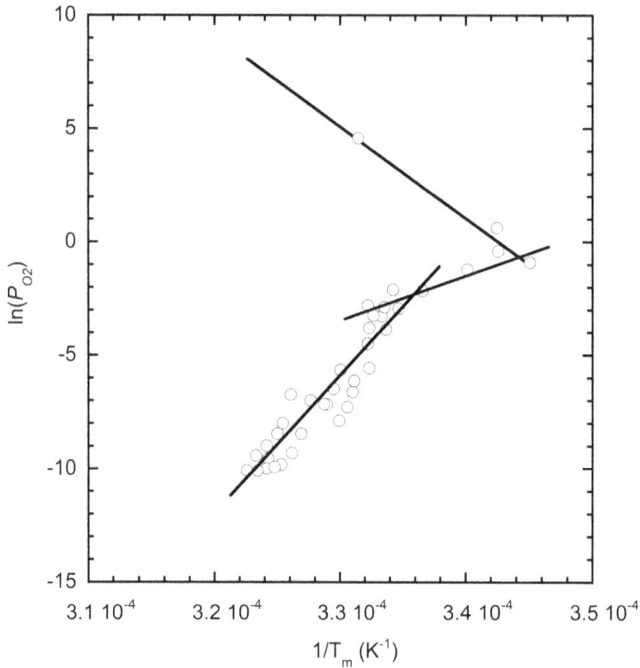

FIGURE 3.12 Relationship between the inverse of solidus temperature and oxygen potential at solidus temperature (Kato et al. [151,152], Kato [165], De Bruycke et al. [166,167]).

solidus temperature was obtained at about 75% Pu in the UO_2–PuO_2 system and increased with decreasing O/M ratio.

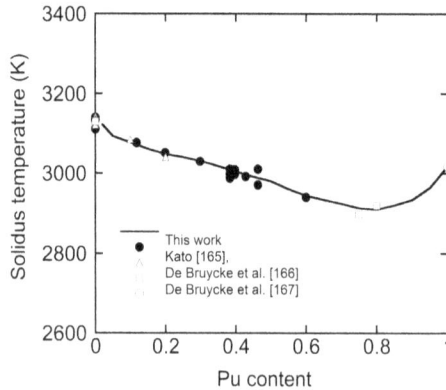

FIGURE 3.13 Solidus temperature in the UO_2–PuO_2 system (Kato et al. [151,152], Kato [165], De Bruycke et al. [166,167]).

The effect of Am and Np content on solidus temperature was evaluated. Figure 3.15 shows the MA content dependence on solidus temperature, and solidus temperature decreased by 6–7 and 1–2 K per 1% increase in Am and Np, respectively.

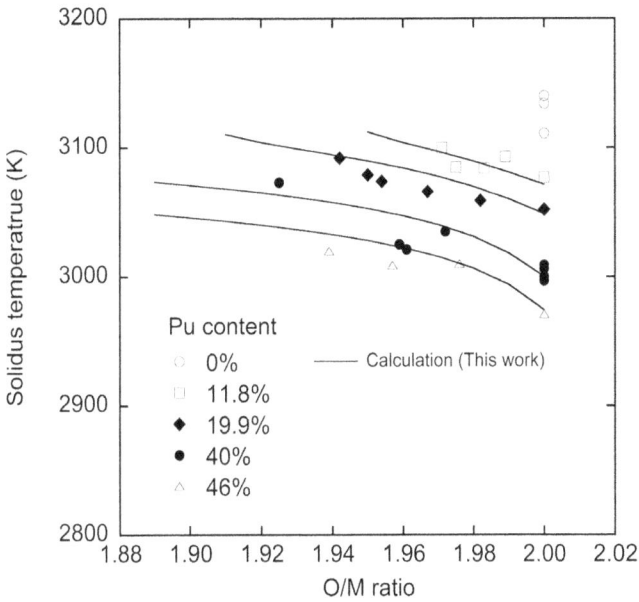

FIGURE 3.14 O/M ratio dependence of solidus temperature (Kato et al. [151, 152], Kato [165]).

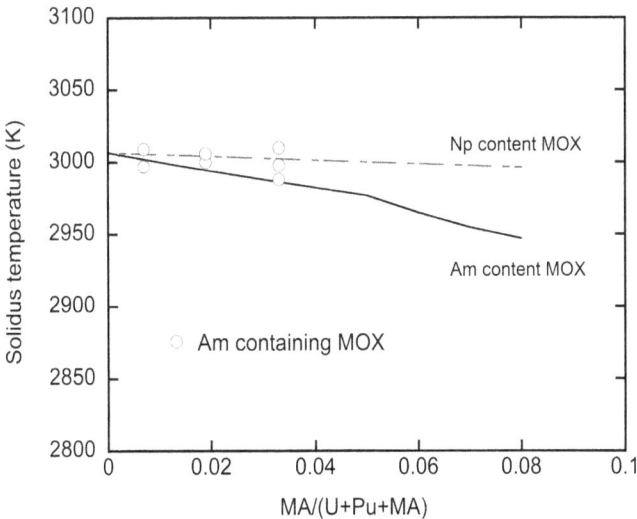

FIGURE 3.15   MA content dependence of the solidus temperature of $(U_{0.61-y}Pu_{0.39}MA_y)O_{2.00}$ (Kato et al. [151,152], Kato [165]).

The calculation results represented the experimental data with a standard deviation of ± 15 K. However, there were no available data on the solidus temperature of hypostoichiometric MOXs with Pu content of more than 50%. The calculation result should be used in the Pu content region of 0%–50%.

3.6 ELECTRICAL CONDUCTION

The electrical conductivity of oxides can be described by electronic and ionic conduction contributions, such as the following equation:

$$\sigma_{\text{total}} = \sigma_{\text{el}} + \sigma_{\text{ion}} \tag{3.56}$$

In oxides, such as CeO_2 and ThO_2, ionic conduction is dominant, which is described by the following equation:

$$\sigma_{\text{ion}} = \frac{C_{\text{ion}}}{T} \cdot \left[V_o^{\cdot\cdot}\right] \cdot \exp\left(-\frac{\Delta_m H_V}{RT}\right) \tag{3.57}$$

The Brouwer diagram of $(U, Pu)O_2$ shows electronic defect dominant as shown in Figure 3.9. This result suggests that electronic conduction contribution is dominant in electrical conduction. Therefore, electrical conduction is represented by the following equation, which is a small polaron hopping conduction mechanism:

$$\sigma_{\text{el}} = \sigma_n + \sigma_p = e[e]\mu_n + e[h]\mu_p = \frac{C_{\mu_e}}{T^{\frac{3}{2}}}[e]\,\exp\left(\frac{E_{\mu_e}}{RT}\right) + \frac{C_{\mu_h}}{T^{\frac{3}{2}}}[h]\,\exp\left(\frac{E_{\mu_h}}{RT}\right). \tag{3.58}$$

where C_{μ_e}, and C_{μ_h} are constants, and E_{μ_e} and E_{μ_h} are electron and hole mobility, respectively.$[e]$, and $[h]$ can be calculated as a function of temperature and P_{O_2} using the Brouwer diagram. The equation has a function of $T^{-\frac{3}{2}}$. Previous works [170,171] reported electrical conduction of UO_2, $(U, Pu)O_2$, and PuO_2. They showed that electronic defects were important in electrical conduction, and 50% Pu had a maximum value in UO_2–PuO_2 solid solution.

Figures 3.16–3.18 show the electrical conduction of UO_2, $(U, Pu)O_2$, and PuO_2 depending on P_{O_2}, respectively. Values of C_{μ_e}, C_{μ_h}, E_{μ_e}, and E_{μ_h} were evaluated by fitting the values with Eq. (3.58) as summarized in Table 3.6. In the figure for UO_2, the experimental data reported by three research groups were plotted [172–174]. All data were measured at almost the same temperature. The data of Lee et al. were lower compared with the other data. The current model could not represent all data, but they were in agreement with the data of Dudney at al. [173] and Matsui and Naito [174].

In the representation of $(U, Pu)O_2$, both E_{μ_e} and E_{μ_h} were determined to be 5 kJ/mol, and C_{μ_e} and C_{μ_h} changed with Pu content. The current model represented the experimental data using the parameters summarized in Table 3.6. The deviation from the experimental data was large in the high P_{O_2} region, and the experimental data were measured at 1,273 K [170,171]. The defects of electron and hole increased with increasing temperature to over 1,500 K. Thus, evaluation using data at temperatures of more than 1,500 K is necessary.

Calculation results of PuO_2 are shown in the figure and well represented the experimental data [175–177], which is shown in Figure 3.18. In a previous work [178],

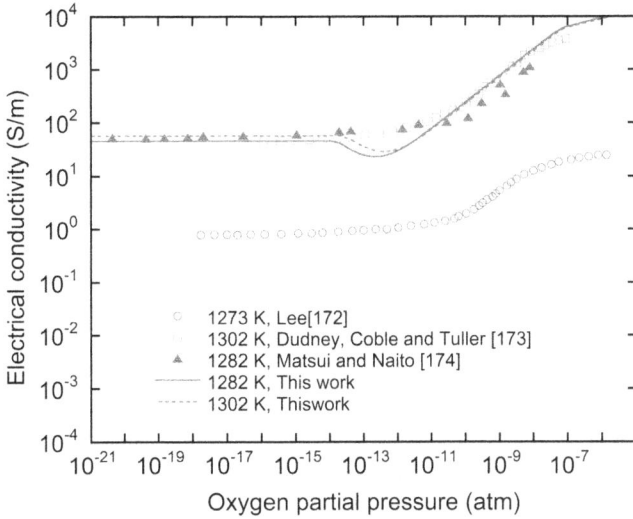

FIGURE 3.16 Electrical conductivity of UO_2 (Lee [172], Dudney, Coble and Tuller [173], Matsui and Naito [174]).

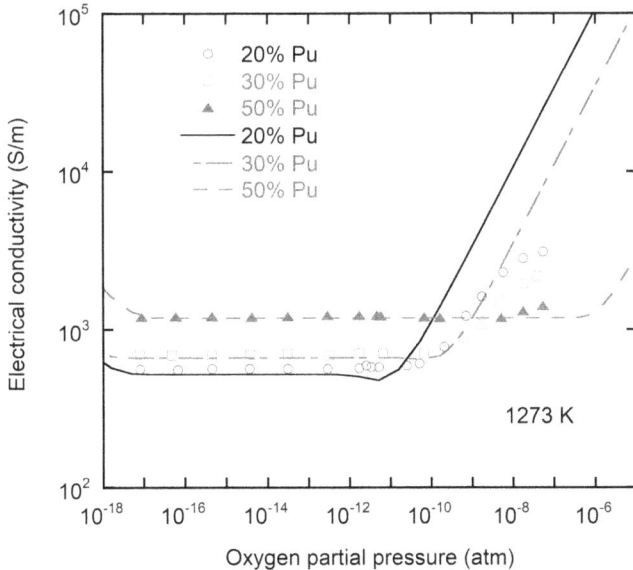

FIGURE 3.17 Electrical conductivity of $(U,Pu)O_2$ (Naito et al. [170], Fujino et al. [171]).

the data of PuO_2 were evaluated using an equation with a function of T^{-1}. Equation (3.58) is more fit to represent the experimental data, which is a function of $T^{-3/2}$. Thus, the defect equilibrium obtained from the oxygen potential measurement data is consistent with the electrical conductivity data.

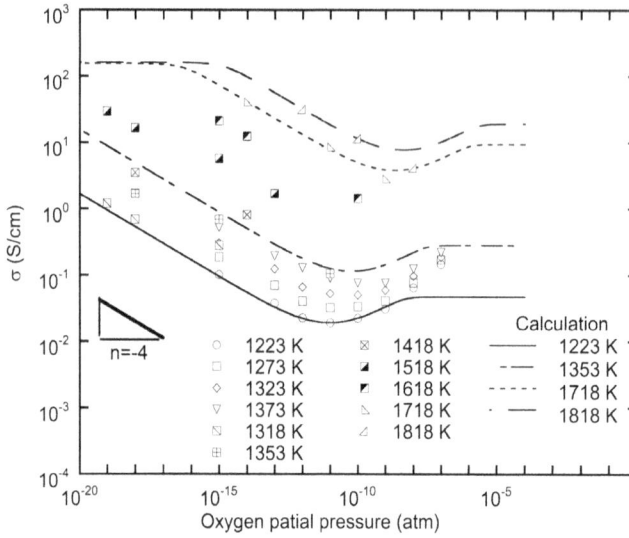

FIGURE 3.18 Electrical conductivity of PuO_2 (Naito et al. [175], Atlas and Schlehman [176], Chereau and Wadier [177].)

TABLE 3.6
Parameters for Eq. (3.48)

	UO_2	$(U_{0.8}Pu_{0.2})O_2$	$(U_{0.7}Pu_{0.3})O_2$	$(U_{0.5}Pu_{0.5})O_2$	PuO_2
C_{μ_e}	1.0E+11	7.00E+10	4.70E+10	3.50E+10	9.0E+7
E_{μ_e}	20,000	5,000	5,000	5,000	30,000
C_{μ_h}	2.0e+8	3.00E+10	3.00E+10	3.50E+10	2.0E+9
E_{μ_h}	20,000	5,000	5,000	5,000	30,000

$C_{\mu_e}, C_{\mu_h}, E_{\mu_e}$ and E_{μ_h}

3.7 SOUND SPEEDS AND MECHANICAL PROPERTIES

Masato Kato and Shun Hirooka

Longitudinal and transverse sound speeds have been measured in pellets with composition and density as parameters. The speed of sound is the underlying data for assessing the velocity of phonons. Mechanical properties, such as Young's modulus obtained from sound speeds, are also used in the evaluation of fuel performance and pellet–cladding mechanical interaction.

Experimental data of the MOX mechanical properties are limited [4,179, 180]. In the study conducted by our group, the longitudinal and transverse sound speeds in (U, Pu)O_{2-x} were measured at room temperature as a function of Pu content, O/M ratio, and density [4]. The sound speed of AmO_2 and NpO_2 was assessed from the

MA content in MOX (Table 3.6). Sound speeds of the MOX were determined using the following equations:

$$V_1 = \left\{ \left(C_U \bullet (V_1)_{UO_2} + C_{Pu} \bullet (V_1)_{PuO_2} + C_{Am} \bullet (V_1)_{AmO_2} + C_{Np} \bullet (V_1)_{NpO_2} \right) \right.$$
$$\left. \bullet (1 - 0.7279 \bullet x) \bullet (1 - 1.3172 \bullet P) \right\} \tag{3.59}$$

$$V_t = \left\{ \left(C_U \bullet (V_t)_{UO_2} + C_{Pu} \bullet (V_t)_{PuO_2} + C_{Am} \bullet (V_t)_{AmO_2} + C_{Np} \bullet (V_t)_{NpO_2} \right) \right.$$
$$\left. \bullet (1 - 1.0545 \bullet x) \bullet (1 - 0.8549 \bullet P) \right\} \tag{3.60}$$

The shear modulus G, Young's modulus E, bulk modulus K, and Poisson's ratio v can be obtained from the sound speeds using the following equations:

$$G = \rho \cdot V_t^2 \tag{3.61}$$

$$E = G \left\{ \frac{\left(3V_1^2 - 4V_t^2 \right)}{\left(V_1^2 - V_t^2 \right)} \right\} \tag{3.62}$$

$$K = \frac{\rho \left(3V_1^2 - 4V_t^2 \right)}{3} \tag{3.63}$$

$$v = \frac{\left(V_1^2 - 2V_t^2 \right)}{2 \left(V_1^2 - V_t^2 \right)} \tag{3.64}$$

The sound speeds and mechanical properties of actinide dioxides at 300 K are summarized in Table 3.7, and E and G changed linearly with Pu content as shown in Figure 3.19 [4,181,182].

Padel and De Novion measured Young's ratio of MOXs using the resonance method [179], and the data are shown in Figure 3.20, where Young's ratio decreased with increasing temperature.

E, G, and K can be represented using the following equations:

$$E = 3 \ K (1 - 2 \cdot v) \tag{3.65}$$

$$G = E / (2 + 2 \cdot v) \tag{3.66}$$

$$K = \gamma \cdot C_1 / (\alpha \cdot V_m) \tag{3.67}$$

γ and v are assumed to be constant regardless of temperature. Mechanical properties can be represented using C_1, α, and V_m, depending on temperature. C_1 representation

TABLE 3.7
Sound Speeds, Mechanical Properties, Debye Temperature, and Grüneisen Constant of Actinide Dioxides

	UO_2	PuO_2	AmO_2	NpO_2
V_l (m/s)	5,358	5,572	4,936	6,500
V_t (m/s)	2,750	2,868	2,343	3,310
G (GPa)	82.2	94.4	64.2	122.0
E (GPa)	219	249	174	323
K (GPa)	204	231	199	308
ν (-)	0.321	0.320	0.355	0.325
T_D (K)	384	406	334	465
γ (-)	1.97	1.97	1.51	3.03

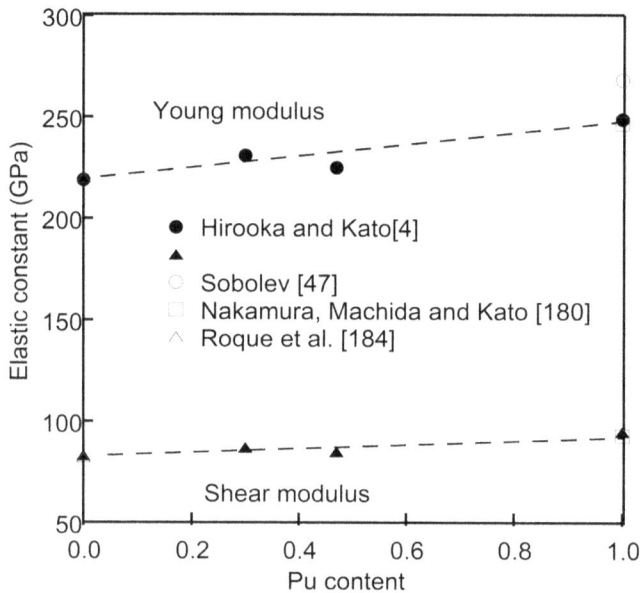

FIGURE 3.19 Pu content dependence of Young's and shear modulus (Hirooka and Kato [4], Sobolev [47], Nakamura, Machida and Kato [179], Roque et al. [182]).

was described in Section 3.8. The calculated Young's ratios of UO_2, MOX, and PuO_2 are shown in Figure 3.20. All data decreased with increasing temperature and were consistent with the experimental data. Hirooka et al [4] calculated Young's ratio as a function of temperature using Debye heat capacity, and similar approach was applied in this study. The calculation results were consistent with the experimental data as shown in Figure 3.20.

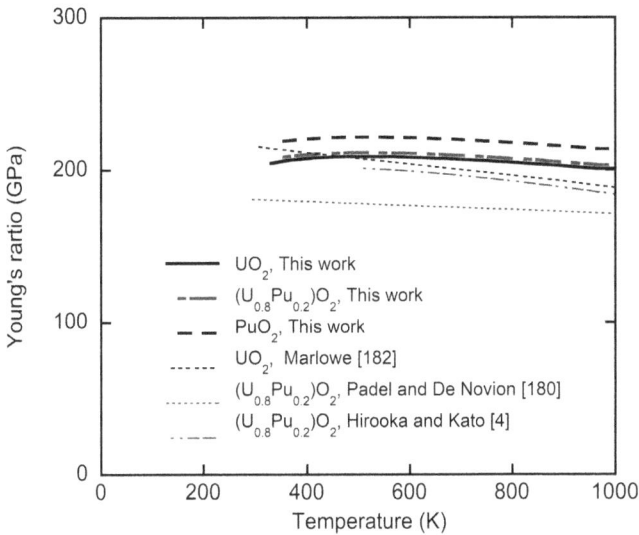

FIGURE 3.20 Temperature dependence of Young' ratios of UO_2, MOX, and PuO_2 (Hirooka and Kato [4], Padel and De Novion [179], Marlowe [183].)

3.8 HEAT CAPACITY

Specific heat C_p of nuclear fuel is a thermal property that is indispensable for performance evaluation and safety evaluation during irradiation. Since various mechanisms are involved in specific heat, it is necessary to understand the dominant mechanism in order to evaluate composition and temperature dependence. Generally, C_p is expressed by the following equation:

$$C_p = C_v + C_d \qquad (3.68)$$

Since C_v approaches $3R$ at high temperatures, it is equivalent to 74.83 J/mol K for a MO_2 molecule having a fluorite structure. Based on the equation, a high value of C_p at high temperatures is caused by C_d. Figure 3.21 compares the C_p of oxides with a fluorite structure. The coefficient of thermal expansion of the oxides is almost the same, which is $2.3 - 3.4 \times 10^{-5}$, which is almost the same. Thus, it is expected that they will take almost the same value of C_p. However, the data are different as shown in the figure. The specific heat values of CeO_2 and ThO_2 are almost the same value and match the data calculated using Eq. (3.68). However, other actinide oxides clearly show higher values compared with CeO_2 and ThO_2. This difference was explained using the Schottky-type specific heat, C_{sch}. Since CeO_2 and ThO_2 do not have f electrons, C_{sch} is not observed, but for other actinide oxides, it is necessary to consider the effect of C_{sch} related to $5f$ electrons.

Comparing the C_p values of UO_2 and PuO_2, PuO_2 shows a value of about 15 J/mol K higher than that of UO_2. This difference is observed by the difference in C_{sch}. The C_p value of MOX is known to be described by Neumann–Kopp's law and is expressed by the composition average of UO_2 and PuO_2. The measurement of C_p in a

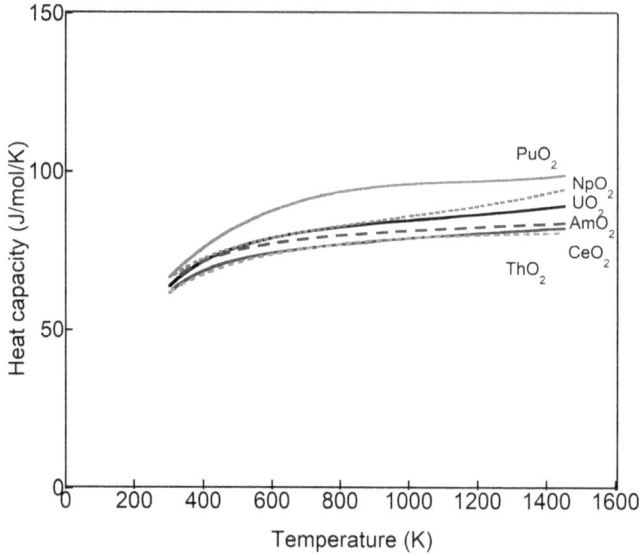

FIGURE 3.21   Heat capacity of oxides having fluorite structure.

temperature range of less than about 1,500 K is mainly measured using differential scanning calorimetry. In higher temperature ranges, drop calorimetry is employed. This method involves measuring the enthalpy change Δh of a sample by dropping it from an arbitrary temperature into a calorimeter. C_p can be obtained using the following equation:

$$C_p = \frac{d\Delta h(T)}{dT} \tag{3.69}$$

Δh is a parameter that is difficult to accurately measure, especially at high temperatures exceeding 2,000 K, because there is a large amount of experimental uncertainty and a slight difference in $\frac{d\Delta h(T)}{dT}$, which have a large effect on C_p. Therefore, the experimental values differ greatly in different studies. As for C_p, the m measurement at high temperatures, results for UO_2 have been reported as shown in Figure 3.22, where the C_p increases in the temperature range exceeding 1,500 K [41,184–188]. This rise is explained by the effects of Frenkel defect and electronic defect contributions, but it is not clear in dominant mechanism. Furthermore, it has been reported that there is a Bredig transition that sharply drops just below the melting point.

As for PuO_2 and MOX, there are almost no measurement data reported at such high temperatures. The C_p of actinide oxides was evaluated assuming Eqs. (3.70) and (3.71).

$$C_p = C_1 + C_d + C_{sch} + C_{exe} \tag{3.70}$$

$$C_v = C_1 + C_{sch} + C_{exe} \tag{3.71}$$

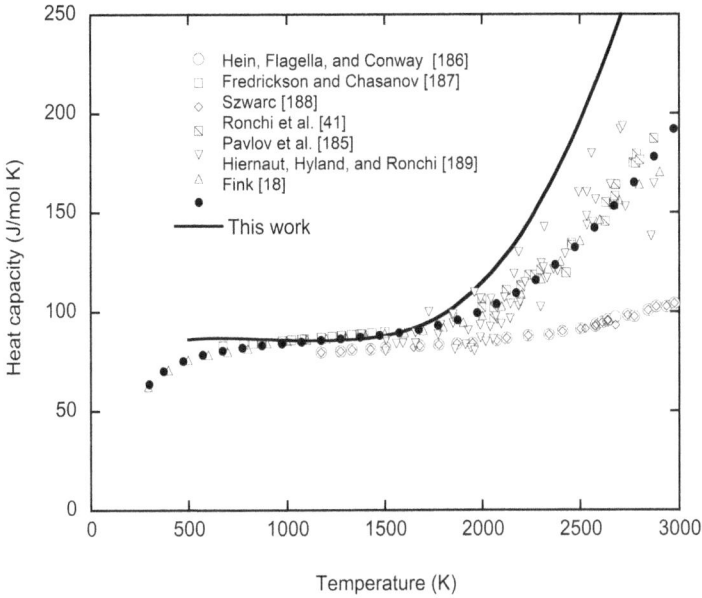

FIGURE 3.22 Heat capacity of UO_2 (Ronchi et al. [41], Pavlov et al. [184], Hein, Flagella, and Conway [185], Fredrickson and Chasanov [186], Fink [18], Szwarc [187], Hiernaut, Hyland, and Ronchi 1993 [188].)

C_1 was calculated using the Debye model:

$$C_1 = 9nR \left(\frac{T}{T_D} \right)^3 \int_0^{T_D/T} \frac{x^4 e^x}{(e^x - 1)^2} dx \tag{3.72}$$

The Debye temperature, T_D, can be written as follows:

$$T_D = \left(\frac{h}{k_B} \right) \left(\frac{9nR}{4 \cdot \pi \cdot a^3} \right)^{\frac{1}{3}} \left(\frac{1}{V_l^3} + \frac{2}{V_t^3} \right)^{-\frac{1}{3}} \tag{3.73}$$

The value for C_d can be obtained using the following equation:

$$C_d = C_v \gamma \alpha T \tag{3.74}$$

The value for γ is obtained at room temperature using the following equation:

$$\gamma = \alpha K V_m / C_v \tag{3.75}$$

Bulk modulus, K, at 300 K was obtained using Eq. (3.63), and γ could be calculated. T_D and γ of actinide oxides are summarized in Table 3.7.

Nakamura and Machida [189] reported the C_{sch} of actinide dioxides was initially obtained using the following equation:

$$C_{sch} = n \cdot R \cdot \frac{\sum_{i>j} \left(\frac{\in_i - \in_j}{T} \right)^2 e^{-\frac{(\in_i - \in_j)}{RT}}}{\left(\sum_i e^{-\frac{\in_i}{RT}} \right)^2} \tag{3.76}$$

Assuming that there are two excited states, Eq. (3.76) is rewritten as follows:

$$C_{sch} = n \cdot R \cdot \frac{\begin{array}{c} \left(\in_1 / RT \right)^2 g_0 \, g_1 e^{-\in_1/RT} + \left(\in_2 / RT \right)^2 g_0 \, g_2 e^{-\in_2/RT} \\ + \left((\in_1 + \in_2)/RT \right)^2 g_1 \, g_2 e^{-(\in_1 + \in_2)/RT} + \left((\in_2 + \in_3)/RT \right)^2 g_2 \, g_3 e^{-(\in_2 + \in_3)/RT} \end{array}}{\left(g_0 + g_1 e^{-\in_1/RT} + g_2 e^{-\in_2/RT} + g_3 e^{-\in_3/RT} \right)^2} \tag{3.77}$$

Table 3.6 summarizes the parameters that are needed to represent the C_{sch} of UO_2, PuO_2, AmO_2, and NpO_2. The C_{sch} of MOXs was determined from the average composition. The parameters were obtained by fitting the initial calculation results using Eq. (3.77).

Two kinds of contributions relating to electron–hole and Frenkel defect pair formation are considered as the heat capacity exited at high temperatures (C_{exe}). According to the Brouwer diagram, electron–hole concentration dominates Frenkel defect concentration. Therefore, the contribution of electron–hole defects is considered as the mechanism of high-temperature specific heat. C_{exe} can be written as follows:

$$C_{exe} = \frac{\Delta H_i^2}{2RT^2} \exp\left(\frac{-\Delta H_i}{2RT} \right) \exp\left(\frac{\Delta S_i}{2R} \right) \tag{3.78}$$

The terms ΔH_i and ΔS_i were evaluated as described in Section 3.3. Figure 3.23 shows the C_p of UO_2, $(U_{0.7}Pu_{0.3})O_{2-x}$, and PuO_2. Konings et al. [190] reported the heat capacity of actinide oxides, but their results are lower than the current data.

In conventional evaluation, the heat capacity of MOXs is described using Kopp's law, which is the composition average of each heat capacity of actinide oxide [43]. Figure 3.23 shows that the heat capacity of MOXs is increased more than that of UO_2 and PuO_2, which means that Kopp's law cannot represent the heat capacity of MOXs at high temperatures. In this study, the heat capacity that exited at high temperatures was expressed by electron–hole pair contribution. In electrical conductivity studies, electronic conduction was dominant, and a small polaron mechanism was proposed in these materials [170–177]. Fujino et al. investigated the electrical conductivity of $(U, Pu)O_{2.00}$ depending on Pu content and showed that the highest conductivity was at around 50% Pu content [171]. Fujino et al. investigated the electrical conductivity of $(U, Pu)O_{2.00}$ depending on Pu content and showed that the highest conductivity was at around 50% Pu content. This report is consistent with the current result as described

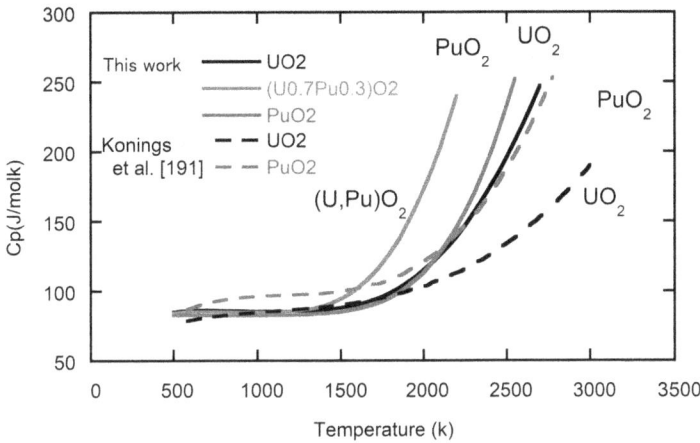

FIGURE 3.23   Heat capacity of UO$_2$, (U$_{0.7}$Pu$_{0.3}$)O$_2$, and PuO$_2$.

in Section 3.6. In the small polaron mechanism, the existence of ions having valences of +5 and +3 is important in the electrical conductivity of actinide dioxides. Ions of U^{5+} and Pu^{3+} are formed in MOXs.

Figure 3.22 shows the heat capacity of UO$_2$ and other data [41,184–188]. The calculated heat capacity values obtained from this work are higher than those of other studies. Experimental data are different among researchers, especially in high-temperature regions. Therefore, it is difficult to determine the correct value. The heat capacities of MOX and PuO$_2$ at high temperatures are limited, and these are indispensable data for evaluation of fuel behavior during transients and accidents. How to evaluate the high-temperature specific heat is a very important issue in this field.

3.9   THERMAL CONDUCTIVITY

Masato Kato, Kyoichi Morimoto, Taku Matsumoto and Keisuke Yokoyama

The temperature of nuclear fuel during burning is a parameter that causes various behaviors and limits the maximum power of the fuel. Therefore, it is important to analyze the fuel temperature during irradiation in the fuel performance evaluation. Thermal conductivity is an indispensable parameter for fuel temperature analysis. Many studies have been conducted since the development of nuclear energy started [10,17,18,32–55,191–199]. However, these studies yield data that have large uncertainties. Recently, the laser-flash method has been employed to measure thermal diffusivity κ and evaluate thermal conductivity λ. Thermal conductivity was determined using the following equation:

$$\lambda = C_p \cdot \rho \cdot \kappa. \tag{3.79}$$

As shown in Eq. (3.79), C_p is required to experimentally evaluate thermal conductivity. Some relational equations to represent C_p have been reported. The equations give different values, especially, in the high-temperature region. In our previous research works, the C_p reported by Carbajo et al. was used in thermal conductivity evaluation [43]. In this work, the experimental data of thermal diffusivity were set in the attached database. Thermal diffusivity of MOXs was listed as a function of Pu content, Np content, Am content, O/M ratio, density, and temperature. It was reevaluated using density and heat capacity calculation results shown in Sections 3.1 and 3.8, respectively.

Through theoretical analysis, it is reported that thermal conductivity is described by phonon and electrical conduction contributions using the following equation:

$$\lambda = \lambda_p + \lambda_e \tag{3.80}$$

Phonon conduction contribution was analyzed using Slack's equations as follows [200]:

$$\lambda_p = 1/(A + BT) \tag{3.81}$$

$$A = \left[\left(\pi^2 V_m T_D \right) / \left(3 h v_p^2 \right) \right] \sum_i \Gamma_i \tag{3.82}$$

$$v_p = (2\pi k_B T_D / h)(V_m / 6\pi^2)^{1/3} \tag{3.83}$$

$$\sum_i \Gamma_i = \Gamma_M + \Gamma_O \tag{3.84}$$

$$\Gamma_i = (1 - x_i) x_i \left[\left(\Delta M_i / M_U \right)^2 + \varepsilon_i \left(\Delta r_i / r_c \right)^2 \right] \tag{3.85}$$

$$\frac{1}{B} = 3.04 \times 10^7 \frac{\bar{M} T_D^3 \delta}{\gamma^2 n^{2/3}} \tag{3.86}$$

All the parameters in Eqs. (3.81)–(3.86) are shown in the Notation table, which can be obtained as a function of composition as shown in previous sections. ε_i was determined to be 100 and 150 for heavy metal and oxygen contribution, respectively, to evaluate composition dependence of λ_p. Small polaron conduction was considered in the electrical conduction mechanism, and the following equation was used:

$$\lambda_e = \left(\frac{\Delta H_i}{e} \right)^2 \cdot \sigma / (4T) = \frac{\Delta H_i^2}{T^{2.5}} \cdot D \cdot K_i^{1/2} \cdot \exp\left(-\frac{\mu}{RT} \right) \tag{3.87}$$

$$D - 0.0868$$

$$\mu = 5 \text{ kJ/mol}$$

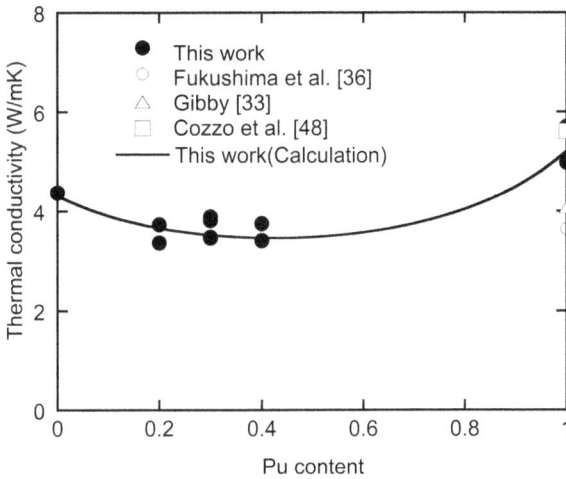

FIGURE 3.24 Pu content dependence of thermal conductivity at 945–1,045 K in the UO₂– PuO₂ system (Fukushima et al. [36], Cozzo et al. [48], Gibby [33].)

Parameters of D and μ were determined to fit the reevaluated data (see Notation table). The following equation was employed to correct the density dependence of thermal conductivity.

$$\lambda = (1-(1-\rho))/(1+0.5*(1-\rho))\,\lambda_{100\%TD} \qquad (3.88)$$

Figure 3.24 shows the Pu content dependence on the thermal conductivity at 950–1,045 K. Thermal conductivity of MOX reached the minimum value at 40%–50% Pu content. The calculated results are presented with a line in Figure 3.24 and are consistent with the experimental data. In this temperature region, the phonon conduction mechanism dominates. The decrease in thermal conductivity was caused by the substitutional solid solution effect between the cations of U and Pu, which act as phonon-scattering factors and increase the parameter A in Eq. (3.81).

In early studies, it was reported that the thermal conductivity of PuO₂ is lower than that of UO₂. However, Cozzo et al. [48] and Matsumoto et al. [197] reported that the thermal conductivity of PuO₂ is higher than that of UO₂. The current analysis showed that the value of PuO₂ is higher as shown in Figure 3.24. In this study, the thermal conductivity of pure material UO₂ and PuO₂ was represented as $1/BT$. Parameter B was obtained using Eq. (3.86), which was calculated from other experimental data, such as sound speed and thermal expansion. There is no doubt about the comparison between the values of UO₂ and PuO₂.

Figures 3.25 and 3.26 show O/M and Am content dependence on thermal conductivity, respectively. The decrease in the O/M ratio and the addition of Am decreased the thermal conductivity in the phonon conduction contribution, which was caused by an increase in the phonon-scattering factor.

Figure 3.27 shows the effect of Am and Np addition on the thermal conductivity of MOX containing 30% Pu at about 1,000 K. The thermal conductivity decreased

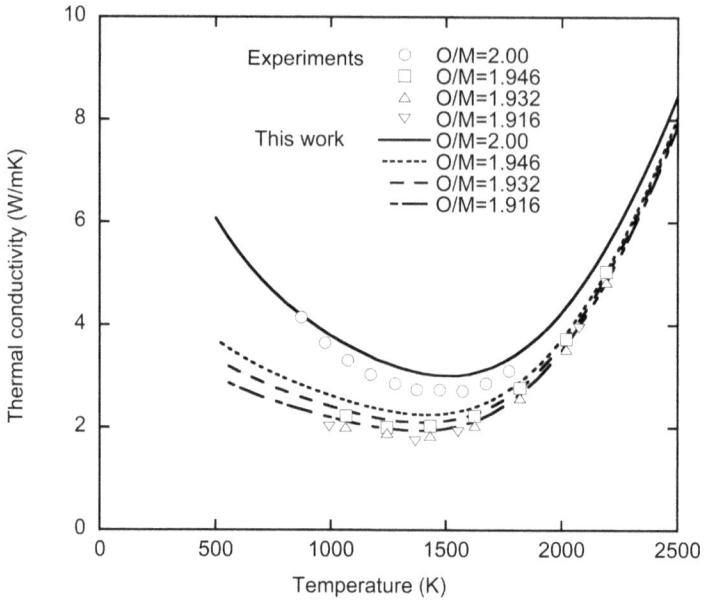

FIGURE 3.25 O/M ratio dependence of the thermal conductivity of MOX containing 30% Pu (100%TD-normalized data).

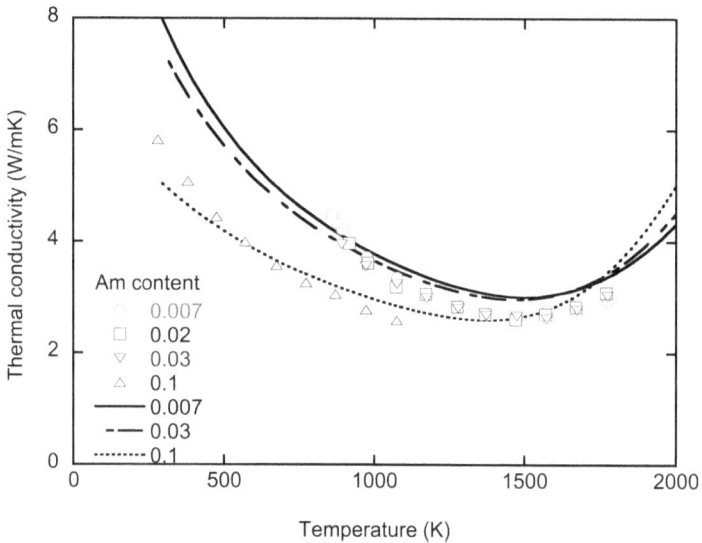

FIGURE 3.26 Am content dependence of the thermal conductivity of MOX containing 30% Pu (100%TD-normalized data).

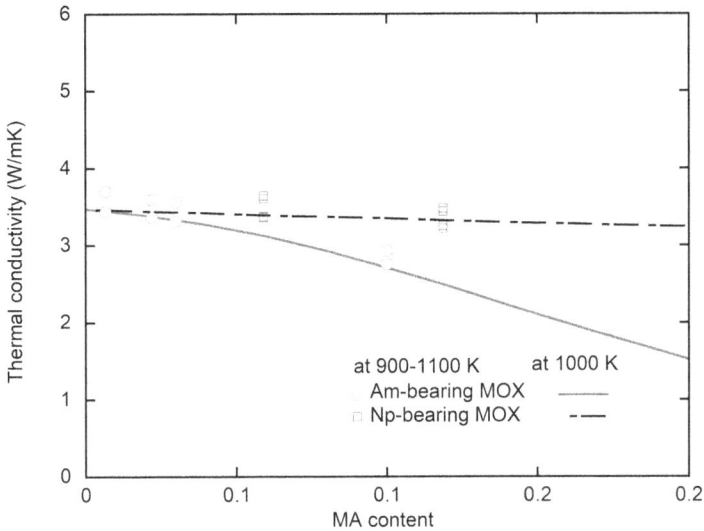

FIGURE 3.27   MA content dependence of the thermal conductivity of MOX containing 30% Pu.

with increasing Am and Np content. The calculation result represented the experimental data. The decrease in thermal conductivity due to Am addition is larger than the decrease in thermal conductivity due to Np addition, which is attributed to the effect of ionic radius and mass on λ_p.

Duriez et al. reported the thermal conductivity of LWR–MOX. They derived a relational correlation between the thermal conductivity of $(U_{0.85}Pu_{0.15})O_{2-x}$ and calculated the data shown in Figure 3.28 [10]. The calculated data from the current work are shown in the figure with a solid line. Both calculation results were consistent at temperatures of less than 1,700 K. Therefore, there is a need to discuss the mechanism of heat capacity and thermal conductivity at higher temperatures. The heat capacity and electrical conductivity data at high temperatures are essential to understand and evaluate the mechanism of thermal conductivity. Bredig transition occurs at high temperatures, and changes in internal energy, point defect concentration, diffusion coefficient, and other properties will occur at the transition point.

3.10   SELF-RADIATION EFFECTS

Masato Kato, Kyoichi Morimoto and Yoshihisa Ikusawa

The properties of compounds containing α emitter elements, such as Pu and Am, change with storage time. This phenomenon is well known as self-irradiated effect. α decay releases α particles and recoil ions, which have energies of about 5 MeV and 90 KeV, respectively. The α particles make several thousands of Frenkel defects, and recoil ions make spike-like defects, which are illustrated in Figure 3.29.

The formation of Frenkel defects causes changes in properties, such as lattice parameter and thermal conductivity. The self-radiation effect of $(U, Pu)O_2$ on lattice

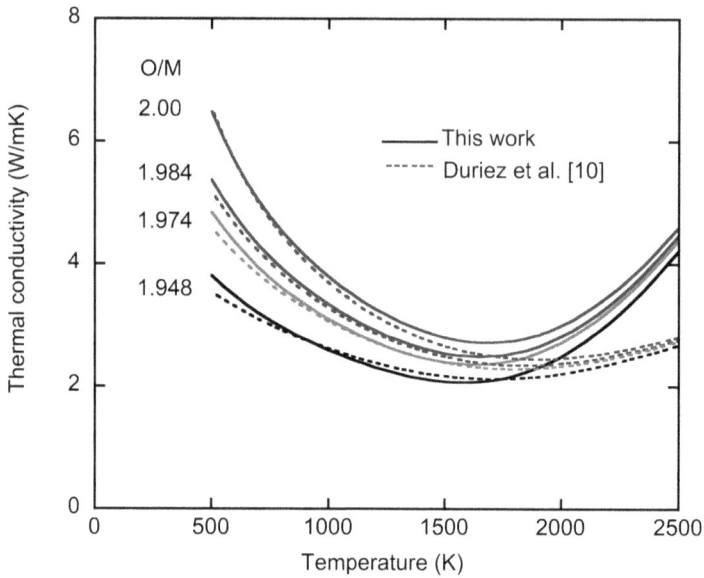

FIGURE 3.28 Thermal conductivity of $(U_{0.85}Pu_{0.15})O_{2-x}$ (Duriez et al. [10]).

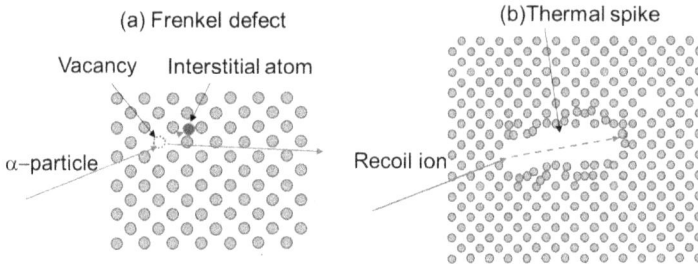

FIGURE 3.29 Defect formation caused by α decay. (a) Frenkel defect and (b) Thermal spike were generated by a particle and recoil ion, respectively.

parameter was investigated in previous work [3], where lattice expansion with storage time and thermal recovery of defects was investigated. Lattice expansion was measured as a function of Pu content and storage time as shown in Figure 3.30, and the following equation was derived:

$$\Delta a/a_0 = 0.0029 \cdot \left(1 - \exp\left(-12,000 \cdot \Lambda' \cdot t\right)\right)_p \tag{3.89}$$

$$\Lambda' = C_{Pu} \sum \Lambda_i \cdot C_i \tag{3.90}$$

Lattice expansion was saturated at 0.29%. There are three kinds of defects, namely, oxygen Frenkel defect, metal Frenkel defect, and defects involving He atom, which

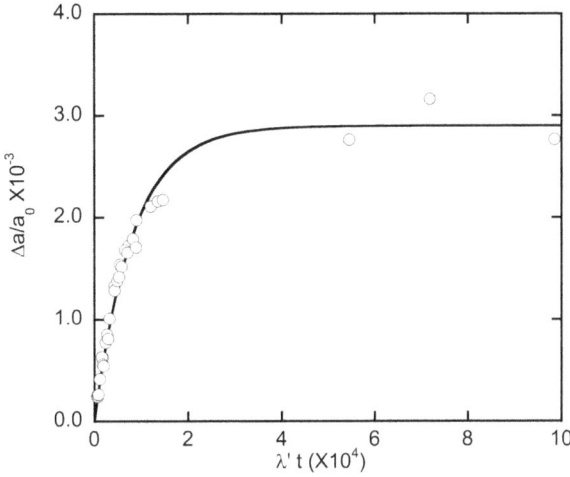

FIGURE 3.30 Lattice expansion of $(U, Pu)O_2$ during storage.

have influenced lattice expansion by 25%, 55%, and 20%, respectively. Thermal recovery kinetics of their defects were determined as follows:

$$K_O = 0.002 \cdot \exp(-1,400/T) \tag{3.91}$$

$$K_M = 0.80 \cdot \exp(-8,500/T) \tag{3.92}$$

$$K_{He} = 5.0 \cdot \exp(-14,000/T) \tag{3.93}$$

Thermal recovery of oxygen Frenkel defect, metal Frenkel defect, and defects involving He atom was observed in temperature regions of less than 673 K, 673–1,073 K, and more than 1,073 K, respectively.

Morimoto et al. [196]. and Ikusawa et al. [201]. reported the thermal recovery of the thermal conductivity of MOXs. Thermal conductivity was degraded by self-irradiation effect, and degraded thermal conductivity was recovered with temperature. The thermal conductivity of MOXs containing 2.5% Pu and 30% Pu are plotted in Figure 3.31. MOX containing 2.5% Pu was measured twice. The data of the second run were completely recovered from the effect of defects.

In this work, Eq. (3.92) was derived to describe thermal conductivity. The self-irradiation effect was considered in the phonon conduction term λ_p. It was assumed that defects yield impurities, and their effect on thermal conductivity was considered in constant A as shown by the following equations:

$$\lambda_p = 1 / (A + A_O + A_M + A_{He} + BT) \tag{3.94}$$

$$A_O = \frac{\exp(-K_O \cdot t - 4.4247)}{0.25} \cdot 0.08 \cdot \frac{\Delta a/a_0}{0.0029} \tag{3.95}$$

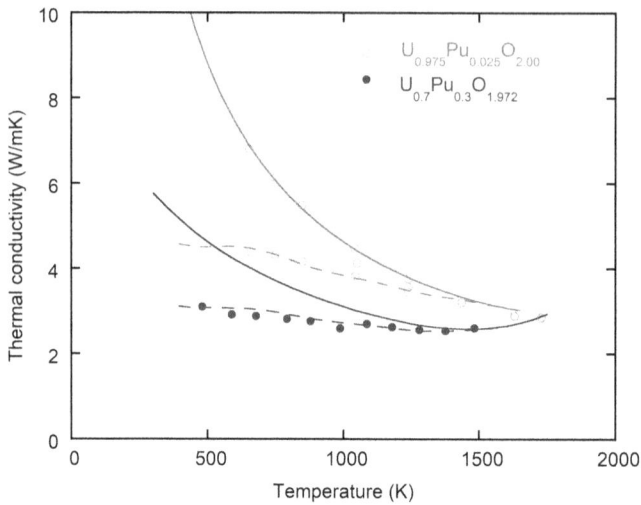

FIGURE 3.31 Thermal conductivity of self-irradiated MOXs. The broken and solid lines indicate self-irradiated and nonirradiated MOXs, respectively.

$$A_M = \frac{\exp(-K_M \cdot t - 4.4687)}{0.55} \cdot 0.03 \cdot \frac{\Delta a/a_0}{0.0029} \qquad (3.96)$$

$$A_{He} = \frac{\exp(-K_{He} \cdot t - 6.4315)}{0.20} \cdot 0.03 \cdot \frac{\Delta a/a_0}{0.0029} \qquad (3.97)$$

Constants of 0.08, 0.03, and 0.03 in each of the above equations are effects of A_O, A_M, and A_{He}, respectively, on λ_p.

Using Eqs. (3.89) and (3.94), thermal recovery of thermal conductivity was evaluated. The calculated results are shown with a line in Figure 3.31. The broken line shows the thermal conductivity recovered with temperature. The figure shows that degraded thermal conductivity was recovered with temperature. The calculation well described the experimental data. In the initial irradiation stage of fresh fuel, the effects of accumulation and recovery of defects on fuel temperature need to be considered in the operation of reactors.

3.11   PHASE SEPARATION

Masato Kato

In the phase diagram of the UO_2–PuO_2–Pu_2O_3 system, it is known that a miscibility gap is observed. Phase separation to two fcc phases, which have different O/M ratios, occurs at a Pu content of over 20% and O/M ratio of less than 2.00 at room temperature. It was considered that phase separation causes the occurrence of microcracks in the production and burning of MOX pellets containing high Pu content. Therefore, phase separation of MOXs has been investigated using microstructure observation, X-ray diffraction, and thermal analysis.

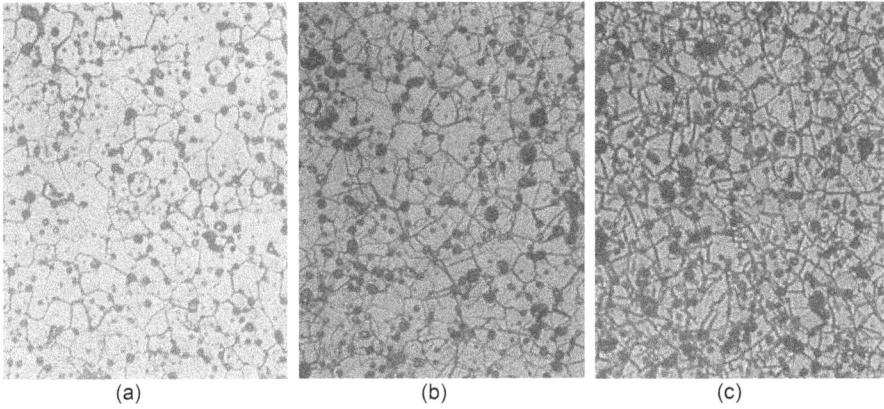

FIGURE 3.32 Microstructures of $(U_{0.7}Pu_{0.3})O_{2-x}$ pellets. with (a) O/M=2.00, (b) O/M=1.944 and (c) O/M=1.932.

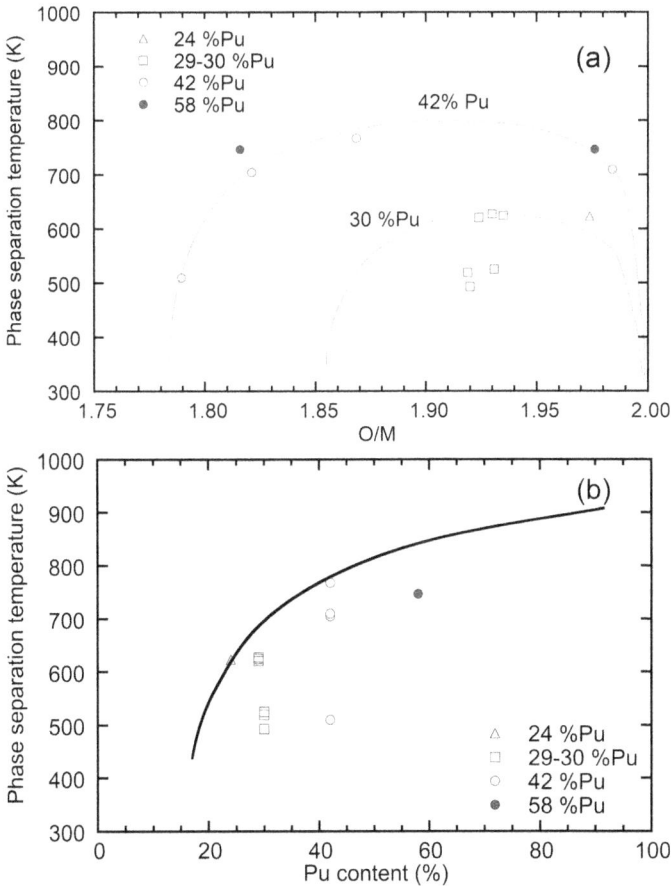

FIGURE 3.33 Phase separation temperature dependence on (a) O/M ratio and (b) Pu content.

Figure 3.32 shows the optical microstructure of $(U_{0.7}Pu_{0.3})O_{2-x}$. The sample with O/M$=2.00$ was an fcc single phase and equiaxial grain. In the samples with O/M$=1.944$ and 1.932, it was observed that phase separation occurred inside the crystal grains.

Figure 3.33 shows the phase separation temperature dependence on the O/M ratio and Pu content. The phase separation temperature of $(U_{0.7}Pu_{0.3})O_{2-x}$ was about 620 K, and two phases of O/M$=1.98$ and 1.889 were observed. The region of phase separation expands in higher Pu content and lower O/M ratio, which is observed like a tunnel. In a number of experiments related to phase separation, occurrence of microcracks is observed in the samples. Phase separation is not considered to affect pellet production and irradiation behavior.

NOTATION

Symbol	Comments	Value, Equation, Unit
G	Diameter of crystal grain at t	μm
G_0	Diameter of crystal grain at $t=0$	μm
K	Equilibrium constant	μm³/h
t	Heat treatment time	h
W	Sample weight of MO$_2$	g
ΔW	Weight change	g
M	Molecular mass of MO$_2$	-
E	Electromotive force	V
F	Faraday constant	96,485 C/mol
$\bar{G}_f$	Standard Gibbs free energy of formation	J/mol
$\Delta \bar{G}_{O_2}$	Oxygen potential	$\Delta \bar{G}_{O_2} = R \cdot T \cdot \ln P_{O_2}$ J/mol
R	Gas constant	8.3145 J/(mol · K)
T	Temperature	K
P_i	Partial pressure of i	atm
C_{Pu}	Pu content	$C_{Pu} = C_{Pu} / (C_U + C_{Pu} + C_{Am} + C_{Np})$
$C_{Pu}{}'$	Equivalent Pu content	$C_{Pu}{}' = C_{Pu} + 2.5 \cdot C_{Am} + 0.5 \cdot C_{Np}$
C_U	U content	$C_U = C_U / (C_U + C_{Pu} + C_{Am} + C_{Np})$
C_{U4+}	U^{4+} content[a]	$C_U = C_{U4+} + C_{U5+}$
C_{U5+}	U^{5+} content[a]	$C_{U5+} = C_{Am}$
C_{Am}	Am content	$C_{Am} = C_{Am} / (C_U + C_{Pu} + C_{Am} + C_{Np})$
C_{Np}	Np content	$C_{Np} = C_{Np} / (C_U + C_{Pu} + C_{Am} + C_{Np})$
a	Lattice parameter	$a = 4/\sqrt{3} \cdot \{r_c (1 + 0.112x) + r_a\}$ Å
r_c	Ionic radius of cation	$r_c = r_{U4+} \cdot C_{U4+} + r_{U5+} \cdot C_{U5+} + r_{Pu} \cdot C_{Pu}$
		$\quad + r_{Am} \cdot C_{Am} + r_{Np} \cdot C_{Np}$, Å
		r_{U4+} : 0.9972
		r_{U5+} : 0.9000
		r_{Pu} : 0.9642
		r_{Am} : 1.050
		r_{Np} : 0.9805
	Ionic radius of O^{2-}	1.372, Å

(Continued)

Symbol	Comments	Value, Equation, Unit
ρ_{th}	Theoretical density	$\rho_{th} = \dfrac{4\bar{M}}{A_v \cdot a^3}$ g/cm$^{.}$
$\bar{M}$	Average molecular mass	g/mol
A_v	Avogadro constant	6.022×10^{23} mol^{-1}
LTE	Linear thermal expansion	$\dfrac{\Delta L}{L_0} = a_0 + a_1 \cdot T + a_2 \cdot T^2 + a_3 \cdot T^3$
a_i	Constant for LTE representation	See Tables 1 and 2
b_i	Constant for LTE representation	See Table 1
x	Deviation in MO$_{2\pm x}$	–
$O_O^\times$	Oxygen atom at lattice point	–
$Vo^{..}$	Oxygen vacancy	–
Oi''	Interstitial oxygen	–
e'	Electron	–
$\left[Vo^{..}\right]$	Concentration of oxygen vacancy	–
$\left[Oi''\right]$	Concentration of interstitial oxygen	–
$\left[e'\right]$	Concentration of electrons	–
$\left[h^{.}\right]$	Concentration of holes	–
K_V	Equilibrium constant of reaction (3.10)	$K_V = \left[Vo^{..}\right]\left[e'\right]^2 P_{O_2}^{1/2}$
K_O	Equilibrium constant of reaction (3.11)	$K_O = \left[Oi''\right]\left[h^{.}\right]^2 P_{O_2}^{-1/2}$
K_i	Equilibrium constant of reaction (3.12)	$K_i = \left[e'\right]\left[h^{.}\right]$
K_F	Equilibrium constant of reaction (3.13)	$K_F = \left[Vo^{..}\right]\left[Oi''\right]$
K_m	Equilibrium constant of reaction m	$K_m = \exp(\Delta S_m / R) \cdot \exp(-\Delta H_m / RT)$
ΔS_m	Entropy of reaction m	–, J/(mol K) for MOX
ΔH_m	Enthalpy of reaction m	–, kJ/mol
ΔS_i	Entropy of reaction (3.12)	85 J/(mol $\cdot$ K) for UO$_2$ $\left(71.1 + 67.35 \cdot C_{Pu} + 108.0 \cdot C_{Pu}^2 - 198.7 \cdot C_{Pu}^3\right),$ J/(mol K) for MOX 105 J/(mol $\cdot$ K) for PuO$_2$
ΔH_i	Enthalpy of reaction (3.12)	300 kJ/mol for UO$_2$ 278 kJ/mol for MOX 350 kJ/mol for PuO$_2$
$h\left(\Delta S_{1/2}\right)_M$	Entropy of $\left(K_{1/2}\right)_M$	$-20.0 - 112.2 \cdot C'_{Pu} + 58.2 \cdot C_{Pu}'^2$, J/(mol K)
$\left(\Delta H_{1/2}\right)_M$	Enthalpy of $\left(K_{1/2}\right)_M$	-105 kJ/mol
$\left(\Delta S_{1/2}\right)_{Pu}$	Entropy of $\left(K_{1/2}\right)_{Pu}$	$262.6 - 276.0 \cdot C'_{Pu}$, J/(mol K).
$\left(\Delta H_{1/2}\right)_{Pu}$	Enthalpy of $K_{1/2}$ $\left(K_{1/2}\right)_{Pu}$	159.3 kJ/mol
$\left(\Delta S_{-1/2}\right)_U$	Entropy of $\left(K_{-1/2}\right)_U$	$42.3 + 48.3 \cdot C'_{Pu} + 110.0 \cdot C_{Pu}'^2 + 99.4 \cdot C_{Pu}'^3$, J/(mol K).
$\left(\Delta H_{-1/2}\right)_U$	Enthalpy of $\left(K_{-1/2}\right)_U$	372 kJ/mol
$\left(\Delta S_{-1/2}\right)_M$	Entropy of $\left(K_{-1/2}\right)_M$	$125.8 - 745.1 \cdot C'_{Pu} + 2678.5 \cdot C_{Pu}'^2 - 3683.4 \cdot C_{Pu}'^3$ $+ 1792.7 \cdot C_{Pu}'^4$, $\cdot$ J/(mol K)
$\left(\Delta H_{-1/2}\right)_M$	Enthalpy of $\left(K_{-1/2}\right)_M$	376.5 kJ/mol

(Continued)

Symbol	Comments	Value, Equation, Unit
$\left(\Delta S_{-1/2}\right)_{Pu}$	Entropy of $\left(K_{-1/2}\right)_{Pu}$	$91.6 - 63.7 \cdot C'_{Pu} - 54.0 \cdot C'^{2}_{Pu} + 99.4 \cdot C'^{3}_{Pu}$, $\cdot J/\left(\text{mol K}\right)$
$\left(\Delta H_{-1/2}\right)_{Pu}$	Enthalpy of $\left(K_{-1/2}\right)_{Pu}$	306 kJ/mol
$\left(\Delta S_{-1/4}\right)_{U}$	Entropy of $\left(K_{-1/4}\right)_{U}$	$72.1 + 82.0 \cdot C'_{Pu} + 164.0 \cdot C'^{2}_{Pu}, \cdot J/\left(\text{mol K}\right)$
$\left(\Delta H_{-1/4}\right)_{U}$	Enthalpy of $\left(K_{-1/4}\right)_{U}$	511 kJ/mol
$\left(\Delta S_{-1/4}\right)_{Pu}$	Entropy of $\left(K_{-1/4}\right)_{Pu}$	$121.4 - 30.0$ J/(molK)
$\left(\Delta H_{-1/4}\right)_{Pu}$	Enthalpy of $\left(K_{-1/4}\right)_{Pu}$	445 kJ/mol
$\Delta S_{-1/3}$	Entropy of $K_{-1/3}$	$275.7 - 1456.5 \cdot C'_{Pu} + 5411.0 \cdot C'^{2}_{Pu} - 7466.2 \cdot C'^{3}_{Pu}$, $+3585.3 \cdot C'^{4}_{Pu}$
$\Delta H_{-1/3}$	Enthalpy of $K_{-1/3}$	$\Delta H_{-1/3} = 892$ kJ/mol
D^{*}_{o}	Oxygen self-diffusion coefficient	Eq. (3.14), m²/s
$\widetilde{D}_{o}$	Oxygen chemical diffusion coefficient	Eq. (3.15), m²/s
D^{0}_{Vo}	Pre-exponential coefficient	3.1×10^{-8} m²/s for MOX
ΔH^{m}_{Vo}	Migration energy of oxygen vacancy	60 kJ/mol for MOX
D^{0}_{Oi}	Pre-exponential coefficient	8.6×10^{-6} m²/s for MOX
ΔH^{m}_{Oi}	Migration energy of interstitial	112.5 kJ/mol for MOX
T_{m}	Solidus temperature	K
σ_{ion}	Ionic conductivity	S/m
C_{ion}	Constant for σ_{ion}	
σ_{el}	Electronic conductivity	S/m
σ_{n}	Electron conductivity	S/m
σ_{p}	Hole conductivity	S/m
$C_{\mu_{e}}$	Constant for σ_{el}	–
$E_{\mu_{e}}$	Migration energy for electron	J/mol
$C_{\mu_{h}}$	Constant for σ_{p}	–
$E_{\mu_{h}}$	Migration energy for hole	J/mol
e	Elementary charge	1.602×10^{-19} C
V_{l}	Longitudinal sound speed of MO_{2-x}	m/s
V_{t}	Transverse sound speed of MO_{2-x}	m/s
$\left(V_{l}\right)_{UO_2}$	Longitudinal sound speed of UO_{2}	5,358 m/s
$\left(V_{l}\right)_{PuO_2}$	Longitudinal sound speed of PuO_{2}	5,572 m/s
$\left(V_{l}\right)_{AmO_2}$	Longitudinal sound speed of AmO_{2}	4,936 m/s
$\left(V_{l}\right)_{NpO_2}$	Longitudinal sound speed of NpO_{2}	6,500 m/s
$\left(V_{t}\right)_{UO_2}$	Transverse sound speed of UO_{2}	2,750 m/s
$\left(V_{t}\right)_{PuO_2}$	Transverse sound speed of PuO_{2}	2,868 m/s
$\left(V_{t}\right)_{AmO_2}$	Transverse sound speed of AmO_{2}	2,343 m/s
$\left(V_{t}\right)_{NpO_2}$	Transverse sound speed of NpO_{2}	3,310 m/s
ρ	Density	g/cm³
p	Porosity	$p = 1 - \dfrac{\rho}{\rho_{th}}$
G	Shear modulus	$G = \rho \cdot V_{t}^{2}$, Pa $G = E/\left(2 + 2 \cdot v\right)$

(*Continued*)

Symbol	Comments	Value, Equation, Unit
E	Young's modulus	$E = G\left\{\dfrac{\left(3V_l^2 - 4V_t^2\right)}{\left(V_l^2 - V_t^2\right)}\right\}$, Pa
		$E = 3\,K\left(1 - 2\cdot v\right)$
K	Bulk modulus	$K = \dfrac{\rho\left(3V_l^2 - 4V_t^2\right)}{3}$, Pa
v	Poisson's ratio	$v = \dfrac{\left(V_l^2 - 2V_t^2\right)}{2\left(V_l^2 - V_t^2\right)}$
T_D	Debye temperature	$T_D = \left(h/k_B\right)\left(9N/4\pi a^3\right)^{1/3}\left(1/V_l^3 + 2/V_t^3\right)^{-1/3}$, K
γ	Grüneisen constant	$\gamma = \alpha K V_m / C_v$,
C_p	Heat capacity at pressure constant	$C_p = C_l + C_d + C_{sch} + C_{exe}$, J/(mol K)
C_l	Lattice heat capacity (Debye model)	$C_l = 9nR\left(\dfrac{T}{T_D}\right)^3 \displaystyle\int_0^{T_D/T} \dfrac{x^4 e^x}{(e^x - 1)^2}\, dx$, J/(mol K)
C_d	Dilatational term of heat capacity	$C_d = C_v \gamma \alpha T$, J/(mol K)
C_{sch}	Schottky-type heat capacity	Eq. (3.58)
C_{exe}	C_p exited at high temperatures	$C_{exe} = \dfrac{\Delta H_i^2}{2RT^2}\exp\left(\dfrac{-\Delta H_i}{2RT}\right)\exp\left(\dfrac{\Delta S_i}{2R}\right)$, J/(mol K)
C_v	Heat capacity at volume constant	$C_l + C_{sch} + C_{exe}$, J/(mol K)
h	Planck constant	1.38×10^{-23} J/K
k_B	Boltzmann constant	6.63×10^{-34} Js
α	Coefficient of thermal expansion	$3 \cdot LTE/\Delta T$
V_m	Atomic volume	$(a \times 10^{-10})^3 / \left(4 \cdot \left(1 + (O/M)\right)\right)$
$\in_i$	ith exited energy	Table 5
g_i	degeneracy of exited state i	Table 5
λ	Thermal conductivity	$\lambda = C_p \cdot \rho \cdot \kappa$ W/(m · K)
κ	Thermal diffusivity	m^2/s
λ_p	Phonon conduction term	$\lambda_s = 1/(A + BT)$ W/(mK)
λ_e	Electronic conduction term	$\lambda_e = \dfrac{\Delta H_i^2}{T^{2.5}} \cdot D \cdot K_i^{1/2} \cdot \exp\left(-\dfrac{\mu}{RT}\right)$, W/(mK)
A	Constant for λ_p	$A = \left[\left(\pi^2 V_m T_D\right)/\left(3h v_p^2\right)\right]\displaystyle\sum_i \Gamma_i$, (mK)/W
B	Constant for λ_p	$\dfrac{1}{B} = 3.04 \times 10^7 \dfrac{\bar{M} T_D^3 \delta}{\gamma^2 n^{2/3}}$, m/W
v_p	Average phonon velocity	$v_p = (2\pi k_B T_D / h)(V_m/6\pi^2)^{1/3}$, m/s
Γ_i	The cross section of element i	$\Gamma_i = (1 - x_i) x_i \left[\left(\Delta M_i / M_U\right)^2 + \varepsilon_i \left(\Delta r_i / r_U\right)^2\right]$
$\bar{M}$	Average mass	$\bar{M} = 238 \cdot C_U + 239 \cdot C_{Pu} + 237 \cdot C_{Np} + 241 \cdot C_{Am} + 16 \cdot (O/M)$
$\bar{M_i}$	Average mass of cation i	$\bar{M_i} = 238 \cdot C_U + 239 \cdot C_{Pu} + 237 \cdot C_{Np} + 241 \cdot C_{Am}$ for Metal
		$\bar{M_i} = 16$ for Oxygen
ΔM_i	Mass change of element i	$M_i - \overline{M_{M^-}}$

(Continued)

Symbol	Comments	Value, Equation, Unit
M_i	Mass of element i	$M_U = 238$, $M_{Pu} = 239$, $M_{Np} = 237$, $M_{Am} = 241$, $M_O = 16 \cdot (O/M)$
Δr_i	Radius change of element i	$r_i - \bar{r}_i$
$\bar{r}_i$	Average radius of elements	$\bar{r}_M = r_{U4+} \cdot C_{U4+} + r_{U5+} \cdot C_{U5+} + r_{Pu} \cdot C_{Pu} + r_{Np} \cdot C_{Np} + r_{Am} \cdot C_{Am}$, Å $\bar{r}_O = (3^{1/2}/(4a) - \bar{r}_M)$, Å
C_O	Oxygen content	$x/2$
ΔM_O	Mass change of anion;	$\Delta M_O = 16$ (O/M)
$\overline{M_O}$	Average mass of anion	16
Δr_O	Change of anion radius	$\Delta r_O = 0.105$
δ	$(V_m)^{1/3}$	$\left[(a \times 10^{-10})^3 / \left(4 \cdot (1 + (O/M)) \right) \right]^{\frac{1}{3}}$
n	The number of atoms per molecule	$n = 1 + (O/M)$
ε_i	Constants	$\varepsilon_M = 90$ $\varepsilon_O = 150$
D	Constants	$D = 0.0868$
μ	Mobility energy of carrier	$\mu = 5$ kJ/mol
Δa	Change of lattice parameter	m
a_0	Lattice parameter at $t = 0$	m
Λ'	Effective decay constant of MOX	$C_{Pu} \sum \Lambda_i \cdot C_i$
t	Time	h
Λ_i	Effective decay constant of C_i	s^{-1}
C_i	Composition of Pu isotope i	–
K_O	Activation energy for oxygen defect	$0.002 \cdot \exp(-1400/T)$
K_M	Activation energy for metal defect	$0.80 \cdot \exp(-8,500/T)$
K_{He}	Activation energy for defect with He	$5.0 \cdot \exp(-14,000/T)$
A_O	Effect of oxygen defect in λ_p	Eq. (3.93)
A_M	Effect of metal defect in λ_p	Eq. (3.94)
A_{He}	Effect of defect with He in λ_p	Eq. (3.95)
Q_e	Activation energy	J/mol
Q_F	Activation energy	J/mol
$Q_{D_{Oi}^*}$	Activation energy	J/mol
$Q_{D_{Vo}^*}$	Activation energy	J/mol
Q_σ	Activation energy	J/mol
$Q_{C_{exe}}$	Activation energy	J/mol
Q_λ	Activation energy	J/mol
$\Delta H_{Enthalpy}$	Enthalpy	$[e] \cdot H_i$
P	Linear heat rate	W/m
Q	power per unit volume	W/m³
T_i	Temperature at r_i	K
ρ_{th}	Radial direction distance	m

[a] This is applied, when $C_{AM} < C_U$ in Am-bearing MOX.

REFERENCES

1. Rao, L. and G. Tian, Recent advances in actinide science, in *the Proceedings of the Eighth Actinide Conference, Actinide 2005, Manchester, UK, July 4-8, 2005*, Alvarez, R., N.D. Bryan and I. May, Eds. 2006. pp. 509–511. RSC Publishing.
2. Kato, M. and K. Konashi, Lattice parameters of (U, Pu, Am, Np)O$_{2-x}$. *Journal of Nuclear Materials*, 2009. **385**(1): pp. 117–121.
3. Kato, M., et al., Self-radiation damage in plutonium and uranium mixed dioxide. *Journal of Nuclear Materials*, 2009. **393**(1): pp. 134–140.
4. Hirooka, S. and M. Kato, Sound speeds in and mechanical properties of (U, Pu)O$_{2-x}$. *Journal of Nuclear Science and Technology*, 2017. **55**(3): pp. 356–362.
5. Markin, T.L. and R.S. Street, The uranium-plutonium-oxygen ternary phase diagram. *Journal of Inorganic and Nuclear Chemistry*, 1967. **29**(9): pp. 2265–2280.
6. Uchida, T., et al., Thermal expansion of PuO$_2$. *Journal of Nuclear Materials*, 2014. **452**(1–3): pp. 281–284.
7. Uchida, T., et al., Thermal properties of UO$_2$ by molecular dynamics simulation. *Progress in Nuclear Science and Technology*, 2011. **2**: pp. 298–602.
8. Kato, M., et al., Thermal expansion measurement and heat capacity evaluation of hypo-stoichiometric PuO$_{2.00}$. *Journal of Nuclear Materials*, 2014. **451**(1–3): pp. 78–81.
9. Kato, M., et al., Thermal expansion measurement of (U, Pu)O$_{2-x}$ in oxygen partial pressure-controlled atmosphere. *Journal of Nuclear Materials*, 2016. **469**: pp. 223–227.
10. Duriez, C., et al., Thermal conductivity of hypostoichiometric low Pu content (U, Pu)O$_{2-x}$ mixed oxide. *Journal of Nuclear Materials*, 2000. **277**(2): pp. 143–158.
11. Vauchy, R., et al., Ceramic processing of uranium–plutonium mixed oxide fuels (U$_{1-y}$Pu$_y$)O$_2$ with high plutonium content. *Ceramics International*, 2014. **40**(7, Part B): pp. 10991–10999.
12. Vauchy, R., et al., Effect of cooling rate on achieving thermodynamic equilibrium in uranium–plutonium mixed oxides. *Journal of Nuclear Materials*, 2016. **469**: pp. 125–132.
13. Vauchy, R., et al., Actinide oxidation state and O/M ratio in hypostoichiometric uranium–plutonium–americium U$_{0.750}$Pu$_{0.246}$Am$_{0.004}$O$_{2-x}$ mixed oxides. *Inorganic Chemistry*, 2016. **55**(5): pp. 2123–2132.
14. Truphémus, T., et al., Structural studies of the phase separation in the UO$_2$–PuO$_2$–Pu$_2$O$_3$ ternary system. *Journal of Nuclear Materials*, 2013. **432**(1–3): pp. 378–387.
15. Besmann, T.M., Modeling the thermochemical behavior of AmO$_{2-x}$. *Journal of Nuclear Materials*, 2010. **402**(1): pp. 25–29.
16. Roth, J., et al., Effects of stoichiometry on thermal expansion of 20 wt percent PuO$_2$-UO$_2$ fast-fast reactor fuel, *Transactions of the American Nuclear Society*, 1967. **10**(2): p. 457
17. Sengupta, A.K., et al., Evaluation of high plutonia (44% PuO$_2$) MOX as a fuel for fast breeder test reactor. *Journal of Nuclear Materials*, 2009. **385**(1): pp. 173–177.
18. Fink, J.K., Thermophysical properties of uranium dioxide. *Journal of Nuclear Materials*, 2000. **279**(1): pp. 1–18.
19. Martin, D.G., The thermal expansion of solid UO$_2$ and (U, Pu) mixed oxides — a review and recommendations. *Journal of Nuclear Materials*, 1988. **152**(2): pp. 94–101.
20. Lorenzelli, R. and M.E.S. Ali, Dilatation thermique d'oxydes mixtes (UPu)O$_{2-x}$ en fonction de l'ecart a la stoechiometrie. *Journal of Nuclear Materials*, 1977. **68**(1): pp. 100–103.
21. Gibby, R.L., Hanford Quarterly Technical Rep, in HEDL-TME 74-3. 1974. p. A–8.
22. Fahey, J.A., R.P. Turcotte, and T.D. Chikalla, Thermal expansion of the actinide dioxides. *Inorganic and Nuclear Chemistry Letters*, 1974. **10**(6): pp. 459–465.
23. Pallmer, B.P.G. and W.F. Sheely, Eds., North West Rep, 1970.
24. Kavdahl, R.E.S. and E.L. Zebroski, General Electric Rep, 1968.
25. Baldock, P.J., W.E. Spindler, and T.W. Baker, The X-ray thermal expansion of near-stoichiometric UO$_2$. *Journal of Nuclear Materials*, 1966. **18**(3): pp. 305–313.
26. Conway, J.B., R.M. Fincel Jr, and R.A. Hein, The thermal expansion and heat capacity of UO$_2$ to 2200 C. *Transactions of the American Nuclear Society*, 1963. **6**(1): p. 153.

27. Leblanc, J.M. and H. Andriessen, EURATOM/USA Repm, 1962.
28. Yamashita, T., et al., Thermal expansions of NpO_2 and some other actinide dioxides. *Journal of Nuclear Materials*, 1997. **245**(1): pp. 72–78.
29. Ferguson, I.F., R.S. Street, and R.W.M. D'Eye, Harwell Rep. 1960, AERE-r.
30. Minato, K., et al., Thermochemical and thermophysical properties of minor actinide compounds. *Journal of Nuclear Materials*, 2009. **389**(1): pp. 23–28.
31. Vauchy, R., A. Joly, and C. Valot, Lattice thermal expansion of $Pu_{1-y}Am_yO_{2-x}$ plutonium–americium mixed oxides. *Journal of Applied Crystallography*, 2017. **50**(6): pp. 1782–1790.
32. Van Craeynest, J.C. and J.C. Weilbacher, Study of the Thermal Conductivity of Mixed Uranium-Plutonium Oxides. CEA-R-3488, 1968, CEA Fontenay-aux-Roses, France.
33. Gibby, R.L., The effect of plutonium content on the thermal conductivity of (U, Pu) O_2 solid solutions. *Journal of Nuclear Materials*, 1971. **38**(2): pp. 163–177.
34. Washington, A.B.G., UK atomic energy authority, 1973.
35. Weilbacher, J.C., Measurement of the Thermal Diffusivity of Mixed Uranium and Plutonium Oxides. Effects of Stoichiometry and Plutonium Content. CEA-R-4572, 1974, CEA Centre d'Etudes Nucleaires de Fontenay-aux-Roses, France.
36. Fukushima, S., et al., Thermal conductivity of near-stoichiometric (U, Pu, Nd)O_2 and (U, Pu, Eu)O_2 solid solutions. *Journal of Nuclear Materials*, 1983. **116**(2): pp. 287–296.
37. Bonnerot, J.M., Thermal Properties of Mixed Uranium and Plutonium Oxides. CEA-R--5450 1988, CEA Centre d'Etudes Nucleaires de Cadarache. France.
38. Harding, J.H. and D.G. Martin, A recommendation for the thermal conductivity of UO_2. *Journal of Nuclear Materials*, 1989. **166**(3): pp. 223–226.
39. Philipponneau, Y., Thermal conductivity of (U, Pu) O_{2-x} mixed oxide fuel. *Journal of Nuclear Materials*, 1992. **188**: pp. 194–197.
40. Lucuta, P.G. and I.J. Hastings, A pragmatic approach to modelling thermal conductivity of irradiated UO_2 fuel: Review and recommendations. *Journal of Nuclear Materials*, 1996. **232**(2–3): pp. 166–180.
41. Ronchi, C., et al., Thermal conductivity of uranium dioxide up to 2900 K from simultaneous measurement of the heat capacity and thermal diffusivity. *Journal of Applied Physics*, 1999. **85**(2): pp. 776–789.
42. Inoue, M., Thermal conductivity of uranium–plutonium oxide fuel for fast reactors. *Journal of Nuclear Materials*, 2000. **282**(2–3): pp. 186–195.
43. Carbajo, J.J., et al., A review of the thermophysical properties of MOX and UO_2 fuels. *Journal of Nuclear Materials*, 2001. **299**(3): pp. 181–198.
44. Kim, D.J., Y.W. Lee, and Y.S. Kim, The thermal conductivity and lattice parameter measurements in a low Ce content $(U_{1-y}Ce_y)O_2$ mixed oxide. *Journal of Nuclear Materials*, 2005. **342**(1–3): pp. 192–196.
45. Nishi, T., et al., Thermal conductivity of neptunium dioxide. *Journal of Nuclear Materials*, 2008. **376**(1): pp. 78–82.
46. Nishi, T., et al., Thermal conductivity of AmO_{2-x}. *Journal of Nuclear Materials*, 2008. **373**(1–3): pp. 295–298.
47. Sobolev, V., Thermophysical properties of NpO_2, AmO_2 and CmO_2. *Journal of Nuclear Materials*, 2009. **389**(1): pp. 45–51.
48. Cozzo, C., et al., Thermal diffusivity and conductivity of thorium–plutonium mixed oxides. *Journal of Nuclear Materials*, 2011. **416**(1–2): pp. 135–141.
49. Ghosh, P.S., et al., Thermal expansion and thermal conductivity of (Th, Ce)O_2 mixed oxides: A molecular dynamics and experimental study. *Journal of Alloys and Compounds*, 2015. **638**: pp. 172–181.
50. Insulander Björk, K. and L. Kekkonen, Thermal–mechanical performance modeling of thorium–plutonium oxide fuel and comparison with on-line irradiation data. *Journal of Nuclear Materials*, 2015. **467**: pp. 876–885.

51. Somayajulu, P.S., et al., Thermal expansion and thermal conductivity of (Th, Pu)O$_2$ mixed oxides: A molecular dynamics and experimental study. *Journal of Alloys and Compounds*, 2016. **664**: pp. 291–303.

52. Epifano, E., et al., High temperature heat capacity of (U, Am)O$_{2\pm x}$. *Journal of Nuclear Materials*, 2017. **494**: pp. 95–102.

53. Vălu, S.O., et al., The high-temperature heat capacity of the (Th, U)O$_2$ and (U, Pu)O$_2$ solid solutions. *Journal of Nuclear Materials*, 2017. **484**: pp. 1–6.

54. Saoudi, M., et al., Thermal diffusivity and conductivity of thorium-uranium mixed oxides. *Journal of Nuclear Materials*, 2018. **500**: pp. 381–388.

55. Rahman, M.J., B. Szpunar, and J.A. Szpunar, Comparison of thermomechanical properties of (U$_x$, Th$_{1-x}$)O$_2$, (U$_x$, Pu$_{1-x}$)O$_2$ and (Pu$_x$, Th$_{1-x}$)O$_2$ systems. *Journal of Nuclear Materials*, 2019. **513**: pp. 8–15.

56. Woodley, R.E., Oxygen potentials of plutonia and urania-plutonia solid solutions. *Journal of Nuclear Materials*, 1981. **96**(1): pp. 5–14.

57. Sørensen, O.T., Thermodynamic studies at higher temperatures of the phase relationships of substoichiometric plutonium and uranium/plutonium oxides, in *International Conference on Plutonium and Other Actinides - Baden-Baden, 10-13 September 1975, North-Holland, Amsterdam*, 1976.

58. Markin, R.L. and E.J. McIver, Thermodynamic and phase studies for plutonium and uranium oxides with application of compatibility calculations, in *Proceedings of the Third International Conference on Plutonium, London, November 22–26, 1965*. Kay, A.E. and M.B. Waldron, Eds. 1965. pp. 845–857. Barnes and Noble, Inc, New≈York.

59. Mari, C.M. and F. Toci, A novel approach to the oxygen activity microdetermination of oxides by EMF measurements. *Journal of the Electrochemical Society*, 1977. **124**(12): pp. 1831–1836.

60. Javed, N.A., Thermodynamic behavior of (U, Pu) mixed-oxide fuels. *Journal of Nuclear Materials*, 1973. **47**(3): pp. 336–344.

61. Chilton, G.R. and J. Edwards, Oxygen potentials of U$_{0.77}$Pu$_{0.23}$O$_{2\pm x}$ in the temperature range 1523–1822K, in *United Kingdom Atomic Energy Authority Northern Division Report*. 1980, United Kingdom Atomic Energy Authority.

62. Rao, P.V., et al., Oxygen potential and thermal conductivity of (U, Pu) mixed oxides. *Journal of Nuclear Materials*, 2006. **348**(3): pp. 329–334.

63. Tetenbaum, M., Thermodynamics of Nuclear Materials 1974. Thermodyn of Nucl Mater, Symp, Proc, Pap and Discuss. Vol. II. 1975, Vienna, Austria.

64. Chilton, G.R. and I.A. Kirkham. The determination of oxygen potentials of hyperstoichiometric U-Pu dioxides in the temperature range 1500–1800K, in *International Conference on Plutonium and other actinides; Baden Baden, Germany, F.R; 10–13 Sep 1975*, Blank, H. and R. Lindner, Eds. 1975. Commission of the European Communities, Karlsruhe (Germany, F.R.), North-Holland, Amsterdam, The Netherlands.

65. Osaka, M., et al., Chemical thermodynamic representation of (U, Pu, Am)O$_{2-x}$. *Journal of Nuclear Materials*, 2005. **344**(1): pp. 230–234.

66. Dawar, R., V. Chandramouli, and S. Anthonysamy, Chemical potential of oxygen in (U, Pu) mixed oxide with Pu/(U+Pu)=0.46. *Journal of Nuclear Materials*, 2016. **473**: pp. 131–135.

67. Swanson, G.C., *Oxygen Potential of Uranium—Plutonium Oxide as Determined by Controlled-Atmosphere Thermogravimetry*. 1976, Los Alamos National Lab. (LANL), Los Alamos, NM.

68. Ackermann, R.J., P.W. Gilles, and R.J. Thorn, High-temperature thermodynamic properties of uranium dioxide. *The Journal of Chemical Physics*, 1956. **25**(6): pp. 1089–1097.

69. Watanabe, M., M. Kato, and T. Sunaoshi, Oxygen potential measurement and point defect chemistry of UO_2. *Transactions of the American Nuclear Society*, 2016. **114**: pp. 1081–1082.

70. Nakamichi, S., M. Kato, and T. Tamura, Influences of Am and Np on oxygen potentials of MOX fuels. *Calphad*, 2011. **35**(4): pp. 648–651.

71. Matsumoto, T., et al., Oxygen potential measurement of $(Pu_{0.928}Am_{0.072})O_{2-x}$ at high temperatures. *Journal of Nuclear Science and Technology*, 2014. **52**(10): pp. 1296–1302.

72. Komeno, A., et al., Oxygen potentials of PuO_{2-x}. *MRS Proceedings*, 2012. **1444**.

73. Kato, M., et al., Oxygen potentials, oxygen diffusion coefficients and defect equilibria of nonstoichiometric (U, Pu)$O_{2\pm x}$. *Journal of Nuclear Materials*, 2017. **487**: pp. 424–432.

74. Kato, M., et al., Oxygen potentials of plutonium and uranium mixed oxide. *Journal of Nuclear Materials*, 2005. **344**(1–3): pp. 235–239.

75. Kato, M., T. Tamura, and K. Konashi, Oxygen potentials of mixed oxide fuels for fast reactors. *Journal of Nuclear Materials*, 2009. **385**(2): pp. 419–423.

76. Kato, M., et al., Oxygen potential of $(U_{0.88}Pu_{0.12})O_{2\pm x}$ and $(U_{0.7}Pu_{0.3})O_{2\pm x}$ at high temperatures of 1673–1873K. *Journal of Nuclear Materials*, 2011. **414**(2): pp. 120–125.

77. Kato, M., et al., Measurement of oxygen potential of $(U_{0.8}Pu_{0.2})O_{2\pm x}$ at 1773 and 1873 K, and its analysis based on point defect chemistry. *Calphad*, 2011. **35**(4): pp. 623–626.

78. Kato, M., K. Konashi, and N. Nakae, Analysis of oxygen potential of $(U_{0.7}Pu_{0.3})O_{2\pm x}$ and $(U_{0.8}Pu_{0.2})O_{2\pm x}$ based on point defect chemistry. *Journal of Nuclear Materials*, 2009. **389**(1): pp. 164–169.

79. Hirooka, S., et al., Relative oxygen potential measurements of (U, Pu)O_2 with Pu = 0.45 and 0.68 and related defect formation energy. *Journal of Nuclear Materials*, 2022. **558**: p. 153375.

80. Kato, M., Oxygen potentials and defect chemistry in nonstoichiometric (U, Pu) O_2, in *Stoichiometry and Materials Science - When Numbers Matter*, Innocenti, A. and N. Kamarulzaman, Eds. 2012. Intech, London, UK.

81. Hirooka, S., et al., Oxygen potential measurement of (U, Pu, Am)$O_{2\pm x}$ and (U, Pu, Am, Np)$O_{2\pm x}$. *Journal of Nuclear Materials*, 2020. **542**: p. 152424.

82. Brouwer, G., A general asymptotic solution of reaction equations common in solid-state chemistry. *Philips Research Reports*, 1954. **9**(5): pp. 366–376.

83. Kröger, F.A. and H.J. Vink, Relations between the concentrations of imperfections in crystalline solids, in *Solid State Physics*, Seitz, F. and D. Turnbull, Eds. 1956. **2**: p. 307. Academic Press, New York.

84. Kofstad, P., *Nonstoichiometry, Diffusion, and Electrical Conductivity in Binary Metal Oxides*. 1972, Wiley-Interscience, New York, US.

85. Kosuge, K., *Chemistry of Non-Stoichiometric Compounds*. 1994. Oxford University Press, Oxford, New York, Tokyo.

86. Watanabe, M., et al., Defect equilibria and thermophysical properties of CeO_{2-x} based on experimental data and density functional theory calculation result. *Journal of the American Ceramic Society*, 2021. **105**(3): pp. 2248–2257.

87. Andersson, D.A., et al., U and Xe transport in $UO_{2\pm x}$: Density functional theory calculations. *Physical Review B*, 2011. **84**(5): p. 054105.

88. Catlow, C.R.A., Point defect and electronic properties of uranium dioxide. *Proceedings of the Royal Society of London. A. Mathematical and Physical Sciences*, 1997. **353**(1675): pp. 533–561.

89. Catlow, C.R.A. and A.B. Lidiard. Theoretical studies of point-defect properties of uranium dioxide, in *Thermodynamics of Nuclear Materials 1974*. 1974. IAEA, Vienna.

90. Clausen, K., et al., Observation of oxygen Frenkel disorder in uranium dioxide above 2000 K by use of neutron-scattering techniques. *Physical Review Letters*, 1984. **52**(14): pp. 1238–1241.

91. Crocombette, J.P., et al., Plane-wave pseudopotential study of point defects in uranium dioxide. *Physical Review B*, 2001. **64**(10): p. 104107.

92. Freyss, M., T. Petit, and J.P. Crocombette, Point defects in uranium dioxide: Ab initio pseudopotential approach in the generalized gradient approximation. *Journal of Nuclear Materials*, 2005. **347**(1–2): pp. 44–51.

93. Gupta, F., G. Brillant, and A. Pasturel, Correlation effects and energetics of point defects in uranium dioxide: A first principle investigation. *Philosophical Magazine*, 2007. **87**(17): pp. 2561–2569.

94. Iwasawa, M., et al., First-principles calculation of point defects in uranium dioxide. *Materials Transactions*, 2006. **47**(11): pp. 2651–2657.

95. Jackson, R.A., et al., The calculation of defect parameters in UO₂. *Philosophical Magazine A*, 1986. **53**(1): pp. 27–50.

96. Konings, R.J.M. and O. Beneš, The heat capacity of NpO₂ at high temperatures: The effect of oxygen Frenkel pair formation. *Journal of Physics and Chemistry of Solids*, 2013. **74**(5): pp. 653–655.

97. Murch, G.E. and C.R.A. Catlow, Oxygen diffusion in UO₂, ThO₂ and PuO₂. A review. *Journal of the Chemical Society, Faraday Transactions 2: Molecular and Chemical Physics*, 1987. **83**(7): pp. 1157–1169.

98. Nerikar, P.V., et al., Energetics of intrinsic point defects in uranium dioxide from electronic-structure calculations. *Journal of Nuclear Materials*, 2009. **384**(1): pp. 61–69.

99. Petit, T., et al., Point defects in uranium dioxide. *Philosophical Magazine B*, 2009. **77**(3): pp. 779–786.

100. Staicu, D., et al., Impact of auto-irradiation on the thermophysical properties of oxide nuclear reactor fuels. *Journal of Nuclear Materials*, 2010. **397**(1–3): pp. 8–18.

101. Terentyev, D., Molecular dynamics study of oxygen transport and thermal properties of mixed oxide fuels. *Computational Materials Science*, 2007. **40**(3): pp. 319–326.

102. Tiwary, P., A. van de Walle, and N. Grønbech-Jensen, *Ab initio* construction of interatomic potentials for uranium dioxide across all interatomic distances. *Physical Review B*, 2009. **80**(17): p. 174302.

103. Vathonne, E., et al., DFT+ U investigation of charged point defects and clusters in UO₂. *Journal of Physics: Condensed Matter*, 2014. **26**(32): p. 325501.

104. Yu, J., R. Devanathan, and W.J. Weber, First-principles study of defects and phase transition in UO₂. *Journal of Physics: Condensed Matter*, 2009. **21**(43): p. 435401.

105. Yun, Y.S. and W.W. Kim. First principle studies on electronic and defect structures of UO₂, ThO₂, and PuO₂, in *Proceedings of the KNS Spring Meeting*. 2007. KNS, Republic of Korea.

106. Cooper, M.W.D., M.J.D. Rushton, and R.W. Grimes, A many-body potential approach to modelling the thermomechanical properties of actinide oxides. *Journal of Physics: Condensed Matter*, 2014. **26**(10): p. 105401.

107. Freyss, M., N. Vergnet, and T. Petit, Ab initio modeling of the behavior of helium and xenon in actinide dioxide nuclear fuels. *Journal of Nuclear Materials*, 2006. **352**(1–3): pp. 144–150.

108. Lu, Y., Y. Yang, and P. Zhang, Charge states of point defects in plutonium oxide: A first-principles study. *Journal of Alloys and Compounds*, 2015. **649**: pp. 544–552.

109. Read, M.S.D., S.R. Walker, and R.A. Jackson, Derivation of enhanced potentials for plutonium dioxide and the calculation of lattice and intrinsic defect properties. *Journal of Nuclear Materials*, 2014. **448**(1–3): pp. 20–25.

110. Tian, X., et al., First principle study of the behavior of helium in plutonium dioxide. *The European Physical Journal B*, 2013. **86**(4): pp. 1–7.

111. Tiwary, P., et al., Interatomic potentials for mixed oxide and advanced nuclear fuels. *Physical Review B*, 2011. **83**(9): p. 094104.

112. Wilhelm, J., *Diffusion in Solids, Liquid, Gases*, 1960. Academic Press, New York.

113. Watanabe, M., T. Sunaoshi, and M. Kato, Oxygen chemical diffusion coefficients of (U, Pu)O$_{2-x}$. *Defect and Diffusion Forum*, 2017. **375**: pp. 84–90.
114. Watanabe, M., M. Kato, and T. Sunaoshi, Oxygen self-diffusion in near stoichiometric (U, Pu)O$_2$ at high temperatures of 1673–1873 K. *Journal of Nuclear Materials*, 2020. **542**: p. 152472.
115. Kato, M., T. Uchida, and T. Sunaoshi, Measurement of oxygen chemical diffusion in PuO$_{2-x}$ and analysis of oxygen diffusion in PuO$_{2-x}$ and (Pu, U)O$_{2-x}$. *Physica Status Solidi (c)*, 2013. **10**(2): pp. 189–192.
116. Kato, M., et al., Oxygen chemical diffusion in hypo-stoichiometric MOX. *Journal of Nuclear Materials*, 2009. **389**(3): pp. 416–419.
117. Ando, K. and Y. Oishi, Diffusion characteristics of actinide oxides. *Journal of Nuclear Science and Technology*, 1983. **20**(12): pp. 973–982.
118. Glasser-Leme, D. and H. Matzke, Dependence upon oxygen potential of the interdiffusion in single crystalline UO$_2$-(U, Pu)O$_2$. *Solid State Ionics*, 1984. **12**: pp. 217–225.
119. Lidiard, A.B., Self-diffusion of uranium in UO$_2$. *Journal of Nuclear Materials*, 1966. **19**(1): pp. 106–108.
120 Marin, J.F. and P. Contamin, Uranium and oxygen self-diffusion in UO$_2$. *Journal of Nuclear Materials*, 1969. **30**(1): pp. 16–25.
121. Matzke, H., Diffusion in doped UO$_2$. *Nuclear Applications*, 1966. **2**(2): pp. 131–137.
122. Matzke, H., Diffusion processes and surface effects in non-stoichiometric nuclear fuel oxides UO$_{2+x}$, and (U, Pu)O$_{2\pm x}$. *Journal of Nuclear Materials*, 1983. **114**(2): pp. 121–135.
123. Nichols, F.A., Transport phenomena in nuclear fuels under severe temperature gradients. *Journal of Nuclear Materials*, 1979. **84**(1): pp. 1–25.
124. Matzke, H., Atomic transport properties in UO$_2$ and mixed oxides (U, Pu)O$_2$. *Journal of the Chemical Society, Faraday Transactions 2: Molecular and Chemical Physics*, 1987. **83**(7): pp. 1121–1142
125. Sari, C., Oxygen chemical diffusion coefficient of uranium-plutonium oxides. *Journal of Nuclear Materials*, 1978. **78**(2): pp. 425–426.
126. Berthinier, C., et al., Thermodynamic assessment of oxygen diffusion in non-stoichiometric UO2 $\pm$ x from experimental data and Frenkel pair modeling. *Journal of Nuclear Materials*, 2013. **433**(1–3): pp. 265–286.
127. Bayoglu, A.S. and R. Lorenzelli, Etude de la diffusion chimique de l'oxygene dans PuO$_{2-x}$ par dilatometrie et thermogravimetrie. *Journal of Nuclear Materials*, 1979. **82**(2): pp. 403–410.
128. Bayoglu, A.S., A. Giordano, and R. Lorenzelli, Mesure de l'autodiffusion de l'oxygene dans PuO$_{2.00}$ par echange isotopique. *Journal of Nuclear Materials*, 1983. **113**(1): pp. 71–74.
129. Bayoglu, A.S. and R. Lorenzelli, Oxygen diffusion in FCC fluorite type nonstoichiometric nuclear oxides MO$_{2\pm x}$. *Solid State Ionics*, 1984. **12**: pp. 53–66.
130. Ando, K., Y. Oishi, and Y. Hidaka, Self-diffusion of oxygen in single crystal thorium oxide. *The Journal of Chemical Physics*, 1976. **65**(7): pp. 2751–2755.
131. Kim, K.C. and D.R. Olander, Oxygen diffusion in UO$_{2-x}$. *Journal of Nuclear Materials*, 1981. **102**(1–2): pp. 192–199.
132. Garcia, P.H., et al., Oxygen diffusion in relation to p-type doping in uranium dioxide. *Journal of Nuclear Materials*, 2010. **400**(2): pp. 112–118.
133. Dorado, B., et al., First-principles calculation and experimental study of oxygen diffusion in uranium dioxide. *Physical Review B*, 2011. **83**(3): p. 035126.
134. Moore, E., C. Guéneau, and J.P. Crocombette, Oxygen diffusion model of the mixed (U, Pu) O$_{2\pm x}$: Assessment and application. *Journal of Nuclear Materials*, 2017. **485**: pp. 216–230.
135. Breitung, W., Oxygen self and chemical diffusion coefficients in UO$_{2\pm x}$. *Journal of Nuclear Materials*, 1978. **74**(1): pp. 10–18.

136. Stan, M. and P. Cristea, Defects and oxygen diffusion in PuO_{2-x}. *Journal of Nuclear Materials*, 2005. **344**(1–3): pp. 213–218.

137. Auskern, A.B. and J. Belle, Oxygen ion self-diffusion in uranium dioxide. *Journal of Nuclear Materials*, 1961. **3**(3): pp. 267–276.

138. Edwards, H.S., A.F. Rosenberg, and J.T. Bittel, *Report No. ASD-TDR-63-635*. Aeronautical Systems Division, Wright-Patterson Air Force Base, OH, 1963.

139. Belle, J., Oxygen and uranium diffusion in uranium dioxide (a review). *Journal of Nuclear Materials*, 1969. **30**(1): pp. 3–15.

140. Dornelas, W. and P. Lacombe, Diffusion sous champ electrique de l'oxygene aux temperatures de 900°A 1000°C dans l'oxyde d'uranium UO_2. *Journal of Nuclear Materials*, 1967. **21**(1): pp. 100–104.

141. Contamin, P., J.J. Bacmann, and J.F. Marin, Autodiffusion de l'oxygene dans le dioxyde d'uranium surstoechiometrique. *Journal of Nuclear Materials*, 1972. **42**(1): pp. 54–64.

142. Deaton, R.L. and C.J. Wiedenheft, The oxygen exchange reaction of PuO_2. *Journal of Inorganic and Nuclear Chemistry*, 1972. **34**(11): pp. 3419–3425.

143. Deaton, R.L. and C.J. Wiedenheft, *Self-Diffusion of Oxygen in ^{238}PuO*. 1973, Monsanto Research Corp., Miamisburg, OH.

144. Murch, G.E. and R.J. Thorn, The mechanism of oxygen diffusion in near stoichiometric uranium dioxide. *Journal of Nuclear Materials*, 1978. **71**(2): pp. 219–226.

145. Vauchy, R., et al., Oxygen self-diffusion in polycrystalline uranium–plutonium mixed oxide $U_{0.55}Pu_{0.45}O_2$. *Journal of Nuclear Materials*, 2015. **467**: pp. 886–893.

146. D'Annucci, F. and C. Sari, Oxygen diffusion in uranium-plutonium oxide fuels at low temperatures. *Journal of Nuclear Materials*, 1977. **68**(3): pp. 357–359.

147. Woodley, R.E. and R.L. Gibby, Room-temperature oxidation of $(U, Pu)O_{2-x}$, in Westinghouse Hanford Company Report, 1973.

148. Suzuki, K., et al., The oxidation rate of $(U_{0.7}Pu_{0.3})O_{2-x}$ with two fcc phases. *Journal of Alloys and Compounds*, 2007. **444–445**: pp. 590–593.

149. Tanaka, K., et al., Oxidation behavior of Am-containing MOX fuel pellets in air. *Energy Procedia*, 2015. **71**: pp. 282–292.

150. Kato, M., et al., The effect of oxygen-to-metal ratio on melting temperature of uranium and plutonium mixed oxide fuel for fast reactor. *Transactions of the Atomic Energy Society of Japan*, 2008. **7**(4): pp. 420–428.

151. Kato, M., et al., Solidus and liquidus temperatures in the UO_2–PuO_2 system. *Journal of Nuclear Materials*, 2008. **373**(1–3): pp. 237–245.

152. Kato, M., et al., Solidus and liquidus of plutonium and uranium mixed oxide. *Journal of Alloys and Compounds*, 2008. **452**(1): pp. 48–53.

153. Pijanowski, S.W. and L.S. DeLuca, *Melting Points in the System $PuO_2 - UO_2$*. 1960, Knolls Atomic Power Laboratory, General Electric Company, Boston, US.

154. Chikalla, T.D., Melting behavior in the system UO_2-PuO_2. *Journal of the American Ceramic Society*, 1963. **46**: p. 323.

155. Latta, R.E. and R.E. Fryxell, Determination of solidus-liquidus temperatures in the UO_{2+x} system ($-0.50<x<0.20$). *Journal of Nuclear Materials*, 1970. **35**(2): pp. 195–210.

156. Adamson, M.G., E.A. Aitken, and R.W. Caputi, Experimental and thermodynamic evaluation of the melting behavior of irradiated oxide fuels. *Journal of Nuclear Materials*, 1985. **130**: pp. 349–365.

157. Lyon, W.L. and W.E. Baily, The solid-liquid phase diagram for the UO_2-PuO_2 system. *Journal of Nuclear Materials*, 1967. **22**(3): pp. 332–339.

158. Yamamoto, K., et al., Melting temperature and thermal conductivity of irradiated mixed oxide fuel. *Journal of Nuclear Materials*, 1993. **204**: pp. 85–92.

159. Tachibana, T., et al., Determination of melting point of mixed-oxide fuel irradiated in fast breeder reactor. *Journal of Nuclear Science and Technology*, 1985. **22**: pp. 155–157.

160. Komatsu, J., T. Tachibana, and K. Konashi, The melting temperature of irradiated oxide fuel. *Journal of Nuclear Materials*, 1988. **54**: p. 38+44.

161. Aitken, E.A. and S.K. Evans, A *Thermodynamic Data Program Involving Plutonium and Uranium at High Temperature.* 1968. Nucleonics Laboratory Quart., General Electric, Pleasanton, California, US, Rep 3, 1968.

162. Konno, K. and T. Hirosawa, Melting temperature of mixed oxide fuels for fast reactors. *Journal of Nuclear Science and Technology*, 2002. **39**: pp. 771–777.

163. Manara, D., et al., Melting of stoichiometric and hyperstoichiometric uranium dioxide. *Journal of Nuclear Materials*, 2005. **342**(1): pp. 148–163.

164. Manara, D., et al., The melting behaviour of oxide nuclear fuels: Effects of the oxygen potential studied by laser heating. *Procedia Chemistry*, 2012. **7**: pp. 505–512.

165. Kato, M., Melting temperatures of oxide fuel for fast reactors, in *International Congress on Advances in Nuclear Power Plants 2009*. 2009. Atomic Energy Society of Japan (AESJ), Tokyo.

166. De Bruycker, F., et al., The melting behaviour of plutonium dioxide: A laser-heating study. *Journal of Nuclear Materials*, 2011. **416**(1): pp. 166–172.

167. De Bruycker, F., et al., On the melting behaviour of uranium/plutonium mixed dioxides with high-Pu content: A laser heating study. *Journal of Nuclear Materials*, 2011. **419**(1): pp. 186–193.

168. Böhler, R., et al., Recent advances in the study of the UO_2–PuO_2 phase diagram at high temperatures. *Journal of Nuclear Materials*, 2014. **448**(1): pp. 330–339.

169. Guéneau, C., et al., Calphad modelling of the U-Pu-Am-O system.pdf, in *INSPYRE*.

170. Naito, K., et al., Electrical conductivity and defect structure of (U, Pu)O_{2+x}. *Journal of Nuclear Materials*, 1989. **169**: pp. 329–335.

171. Fujino, T., et al., High temperature electrical conductivity and conduction mechanism of (U, Pu)$O_{2\pm x}$ at low oxygen partial pressures. *Journal of Nuclear Materials*, 1993. **202**(1): pp. 154–162.

172. Lee, H.M., Electrical conductivity of UO_{2+x}. *Journal of Nuclear Materials*, 1974. **50**(1): pp. 25–30.

173. Dudney, N.J., R.L. Coble, and H.L. Tuller, Electrical conductivity of pure and yttria-doped uranium dioxide. *Journal of the American Ceramic Society*, 1981. **64**(11): pp. 627–631.

174. Matsui, T. and K. Naito, Electrical conductivity measurement and thermogravimetric study of lanthanum-doped uranium dioxide. *Journal of Nuclear Materials*, 1986. **138**(1): pp. 19–26.

175. Naito, K., et al., Electrical conductivity anomaly in near-stoichiometric plutonium dioxide. *Journal of Nuclear Materials*, 1980. **95**(1): pp. 181–184.

176. Atlas, L.M. and G.J. Schlehman, *Defect Equilibria of PuO_{2-x}, 1100 to 1600°C; Equilibres de Defauts de PuO_{2-x}, dans;Intervalle Compris Entre 1100 et 1600°C; Defektnoe ravnovesie PuO_{2-x} diapazone temperatur 1100–1600°C; Equilibrio de Defectos en PuO_{2-x}, Entre 1100 to 1600°C.* 1966. International Atomic Energy Agency (IAEA), Vienna.

177. Chereau, P. and J.F. Wadier, Mesures de resistivite et de cinetique d'oxydation dans PuO_{2-x}. *Journal of Nuclear Materials*, 1973. **46**(1): pp. 1–8.

178. Kato, M., et al., Defect chemistry and basic properties of non-stoichiometric PuO_2. *Defect and Diffusion Forum*, 2017. **375**: pp. 57–70.

179. Padel, A. and C.H. De Novion, Constantes elastiques des carbures, nitrures et oxydes d'uranium et de plutonium. *Journal of Nuclear Materials*, 1969. **33**(1): pp. 40–51.

180. Nutt Jr, A.W., A.W. Allen, and J.H. Handwerk, Elastic and anelastic response of polycrystalline UO_2-PuO_2. *Journal of the American Ceramic Society*, 1970. **53**(4): pp. 205–210.

181. Nakamura, H., M. Machida, and M. Kato, LDA+U study on plutonium dioxide with spin-orbit couplings. *Progress in Nuclear Science and Technology*, 2011. **12**: pp. 16–19.

182. Roque, V., et al., Effects of the porosity in uranium dioxide on microacoustic and elastic properties. *Journal of Nuclear Materials*, 2000. **277**: pp. 211–216.

183. Marlowe, M.O., High temperature isothermal elastic moduli of UO_2. *Journal of Nuclear Materials*, 1969. **33**(2): pp. 242–244.

184. Pavlov, T.R., et al., Measurement and interpretation of the thermo-physical properties of UO_2 at high temperatures: The viral effect of oxygen defects. *Acta Materialia*, 2017. **139**: pp. 138–154.

185. Hein, R.A., P.N. Flagella, and J.B. Conway, High-temperature enthalpy and heat of fusion of UO_2. *Journal of the American Ceramic Society*, 1968. **51**(5): pp. 291–292.

186. Fredrickson, D.R. and M.G. Chasanov, Enthalpy of uranium dioxide and sapphire to 1500 K by drop calorimetry. *Journal of Chemical Thermodynamics*, 1970. **2**(5): pp. 623–629.

187. Szwarc, R., The defect contribution to the excess enthalpy of uranium dioxide-calculation of the frenkel energy. *Journal of Physics and Chemistry of Solids*, 1969. **30**: pp. 705–711.

188. Hiernaut, J. P., G. J. Hyland and C. Ronchi, Premelting transition in uranium dioxide, *International Journal of Thermophysics*, 1993. **4**(2): pp. 259–283.

189. Nakamura, H. and M. Machida, Numerical calculations for heat capacity of actinide dioxides, in *23th International Conference on Nuclear Engineering ICONE23*. 2015. Japan Society of Mechanical Engineers, Chiba, Tokyo.

190. Konings, R.J.M., et al., The thermodynamic properties of the *f*-elements and their compounds. Part 2. The lanthanide and actinide oxides. *Journal of Physical and Chemical Reference Data*, 2014. **43**(1): p. 013101.

191. Yokoyama, K., et al., Measurements of thermal conductivity for near stoichiometric ($U_{0.7-z}Pu_{0.3}Am_z)O_2$ (z = 0.05, 0.10, and 0.15). *Nuclear Materials and Energy*, 2022. **31**: 101156.

192. Morimoto, K., et al., Thermal conductivities of (U, Pu, Am)O_2 solid solutions. *Journal of Alloys and Compounds*, 2008. **452**(1): pp. 54–60.

193. Morimoto, K., et al., Thermal conductivity of (U, Pu, Np)O_2 solid solutions. *Journal of Nuclear Materials*, 2009. **389**(1): pp. 179–185.

194. Morimoto, K., et al., Thermal conductivities of hypostoichiometric (U, Pu, Am)O_{2-x} oxide. *Journal of Nuclear Materials*, 2008. **374**(3): pp. 378–385.

195. Morimoto, K., M. Kato, and M. Ogasawara, Thermal diffusivity measurement of (U, Pu)O_{2-x} at high temperatures up to 2190K. *Journal of Nuclear Materials*, 2013. **443**(1–3): pp. 286–290.

196. Morimoto, K., et al., Recovery behaviours of thermal conductivities in self-irradiated MOX fuel, in *IOP Conference Series: Materials Science and Engineering*, 2010. **9**: p. 012008. IOP Publishing, Bristol, UK.

197. Matsumoto, T., et al., Thermal conductivity measurement of ($Pu_{1-x}Am_x)O_2$ (x = 0.028, 0.072). *Journal of Alloys and Compounds*, 2015. **629**: pp. 92–97.

198. Kato, M., et al., Physical properties and irradiation behavior analysis of Np- and Am-bearing MOX fuels. *Journal of Nuclear Science and Technology*, 2011. **48**: pp. 646–653.

199. Morimoto, K., et al., The influence of Pu-content on thermal conductivities of (U, Pu) O_2 solid solution, in *Proceedings of International Conference on Fast Reactors and Related Fuel Cycles 2009*. 2009. *Kyoto, Japan*. International Atomic Energy Agency.

200. Slack, G.A., The thermal conductivity of nonmetallic crystals. *Solid State Physics*, 1979. **34**: pp. 1–71.

201. Ikusawa, Y., et al., The effects of plutonium content and self-irradiation on thermal conductivity of mixed oxide fuel. *Nuclear Technology*, 2018. **205**(3): pp. 474–485.

4 Theoretical and Computational Works on Oxide Nuclear Fuel Materials

Masahiko Machida
Japan Atomic Energy Agency

CONTENTS

DOI: 10.1201/9781003298205-4

This chapter briefly reviews theoretical and computational works on nuclear fuel oxide materials. This review is not a complete one overviewing the whole of the field but a biased one by the author's limited experience and knowledge. However, we believe that it is interesting for students and researchers who want to immediately catch the trend of theoretical and computational works on nuclear fuel materials. Especially, we pile up our notions in density functional theory (DFT) calculations on oxide fuel compounds. The readers can easily learn why DFT approaches are important but difficult on the compounds. Furthermore, we show a brand new methodological technique called machine-learning molecular dynamics (MLMD) automatically bridging the time and space gap between DFT and molecular dynamics (MD) in comparison with conventional MD using empirical potentials. We believe that MLMD will be a central scheme in the future. In addition to our specialties, DFT and MD topics, we briefly give an introduction on physics, chemistry, and multiscale simulations with material behaviors on oxide fuel materials through our learning from several literature. We believe that the references in each section will also be useful for students and young researchers.

4.1 OVERVIEW

The ultimate goal of R&D on nuclear fuel materials is to construct a model that captures multiscale fuel behaviors in reactor operating and severe accident conditions. The obtained model should be useful not only to fuel researchers, but also to nuclear reactor engineers and designers.

Recently, a modeling approach grounded on a microscopic basis without empirical modeling as far as possible (i.e., so-called multiscale modeling) has been intensively investigated in the nuclear fuel research community. The trend is very attractive because of its potential predictability, but the road to the goal is not easy due to the rich complexities exhibited by fuel materials. The variety in research issues ranges from the thermal behaviors in single crystals to those in highly damaged polycrystalline, including several types of defects, impurities, and fission gas atoms, among others.

In fuel material research, the most fundamental studies include rather accurate approaches based on electronic structure calculations in single crystals with heavy computational costs. These can be assigned as the so-called benchmark studies using the most advanced methodology on electronic structure calculations. On the contrary, because real fuel materials are basically damaged polycrystalline, a crucial point for understanding rich complexities is modeling consistency in the presence of various types of defects and fission gas migration, under damaged microstructure evolution, swelling, hardening, etc. Then the modeling becomes naturally multiscale and demands sufficient knowledge and experimental assistance.

The purpose of this chapter is not to review the entirety of the multiscale modeling on fuel materials, but to introduce theoretical and computational schemes for the material properties of oxide fuel materials from the UO_2 employed in the present light-water reactors (LWRs) to other actinide and mixed oxides (MOXs). We mainly concentrate herein on the fundamental topic and only briefly touch on the multiscale concepts and approaches. The reason is partly due to the author's research experiences. The author's theoretical group is limited to a few members; hence, they cannot fully cover the multiscale behaviors and modeling of nuclear fuel materials. The other reason is the page limitation, because there is a huge amount of work on multiscale schemes related to their rich complex behaviors.

Thus, this chapter mainly describes the electronic structure calculations and molecular dynamics (MD) studies on oxide fuel materials and briefly introduces multiscale computational works. Please see the recent review papers for the latter multiscale issues [1,2]. Several good reviews and systematic papers on the former fundamental studies are also available, as seen in Refs. [3,4]. We note our incompleteness as a report reviewing the rich fields but introduce one of the most advanced approaches. The approach uses machine-learning techniques to create complicated force fields on atoms from electronic structure calculations. We call this MLMD. The MLMD is a promising method of performing a multiscale simulation that automatically bridges the gap between electronic structural calculations and MD approaches.

The contents of this chapter are presented in seven sections. Section 4.1 provides an overview of the chapter. Section 4.2 discusses the electronic structure calculations and some related remarks. Section 4.3 describes the MD with interatomic potentials composed of empirical and machine learning. Section 4.4 briefly introduces the mesoscale schemes and their varieties. Section 4.5 summarizes the high-temperature thermodynamical behaviors using computational works. Subsequently, Section 4.6 shows the defect chemistry and related high-temperature dynamics. Section 4.7 presents a short introduction of the concept of multiscale modeling, which is the recent trend in the research community of nuclear fuel materials.

4.2 DFT CALCULATIONS AND SOME RELATED REMARKS

Hiroki Nakamura and Masahiko Machida

As mentioned briefly in Section 4.1, an essential target in nuclear fuel modeling is the analysis of the fuel performance and safety that are strongly relevant to the thermophysical behaviors of fuel materials. Modeling enables us to predict thermal transport efficacy. This is of great importance because of the direct relevance of the nuclear energy supply and assessment on the margin to the failure conditions driven by the critical evolutional changes of the material properties in an extremely high-temperature range.

Nuclear fuel modeling historically started from being somewhat empirical (i.e., by fitting the experimental data on nuclear fuel materials); therefore, its deployment was limited to the experimentally accessible range within the measurement accuracy. Attempts to pass over the limitation were then made using physics-based non-empirical simulations, such as microscopic electronic structure calculations and MD simulations. Subsequently, information transfer to meso- and macroscale-level modeling from the abovementioned microscopic models was developed. In this section, we focus on the most fundamental approach grounded on a non-empirical basis (i.e., first-principle calculations on the electronic structures of oxide fuel materials), which is called the "ab-initio" calculations. Almost all works on electronic structural calculations use the DFT because it is one of the most successful ideas in fundamental physics related to solid-state materials.

We also stress that the DFT methodology recently became applicable to the direct estimation of the thermal transport, which is one of the most essential features in fuel materials, by combining the DFT calculation results with the Boltzmann equations for phonons and other quasi-particle transports [5]. These approaches are greatly attractive because they enable the assessment of thermal transports on target materials in "ab-initio" levels, that is, in no empirical ways. Furthermore, the DFT approach is useful for use in studies on defect chemistry and related issues, which are rather important for materials exposed to strong radiation fields. These issues will be briefly discussed in Section 4.6.

So far, DFT has been developed to calculate electronic structures and predict associated physical properties of various materials by using local density approximation (LDA) and generalized gradient approximation (GGA). However, the DFT often fails to predict the electronic states of strongly correlated electron systems. Unfortunately, UO_2 and other actinide oxides belong to such unsuccessful systems. One exception, however, is ThO_2, in which strongly correlated f-electron bands are fortunately unoccupied by transferring all valence electrons into oxide atoms. These failures in UO_2 and other oxides are caused by the LDA and the GGA tending to delocalize electrons, which mainly results in a self-interaction error for the localized features intrinsic to the f-electrons. Actinide dioxides, UO_2, and others are basically insulators with relatively large gaps that are classified into strongly correlated materials. In these materials, LDA and GGA are not that suitable and predict false metallic electronic structures. Therefore, special treatments have been required to consider the strong correlation effects to reproduce the electronic states of AnO_2 when using the DFT. This section discusses the calculations of the electronic states of AnO_2 using the DFT by considering the strong correlations of the f-orbital electrons. First, we will explain the expected electronic ground states of AnO_2. Next, we will introduce DFT calculation methods to account for the strongly correlated features. Afterward, we will review the calculation results for AnO_2 in the remaining part. For a more detailed information about the electronic states of AnO_2 and their DFT calculations, the authors recommend the good review papers as Refs. [4,6,7].

4.2.1 GROUND STATES OF ACTINIDE DIOXIDES

AnO_2 is an insulator composed of a tetravalent cation An^{4+} and a divalent anion O^{2-}. The outermost electrons of An ions are in 5f orbitals and often form open shells, where the electrons are localized due to the strong electronic correlations. The number of f-electrons in the 5f orbitals depends on the An ion species (e.g., 2 for U^{4+} and 4 for Pu^{4+}).

AnO_2 mostly has a fluorite crystal structure (Figure 4.1) that belongs to the $Fm\bar{3}m$ space group. The degeneracy of the f-electrons tends to be relatively large due to their highly symmetric cubic structure. In addition, spin–orbit coupling (SOC), as a relativistic effect, is not negligible due to the 5f-orbitals with a high angular momentum. Consequently, the f-orbitals of AnO_2 are split into 6-degenerate $j = 5/2$ and 8-degenerate $j = 7/2$ orbitals and further split by a crystal field (Figure 4.2a). Finally, the lowest level becomes 4-degenerate Γ_8, and the first excited level is 2-degenerate Γ_7. Focusing on the An ions up to Cm, the f-electrons of ThO_2, PuO_2, and CmO_2 form closed shells (Figure 4.2b); therefore, any magnetic dipole or multipole order is

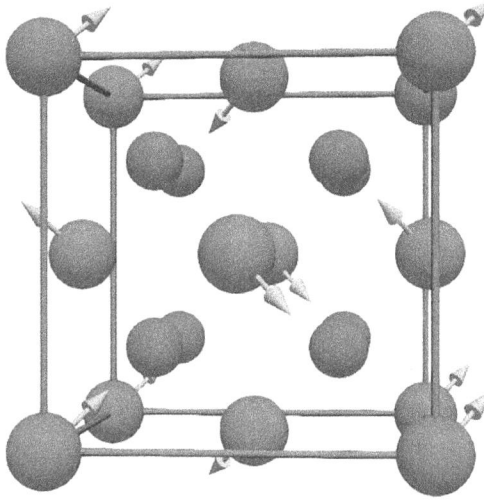

FIGURE 4.1 Typical crystal structure of AnO_2. The large (red) and small (blue) spheres correspond to the An and O ions, respectively. The arrows show the directions of the magnetic moments in the transverse triple-Q antiferromagnetic order in the UO_2 case.

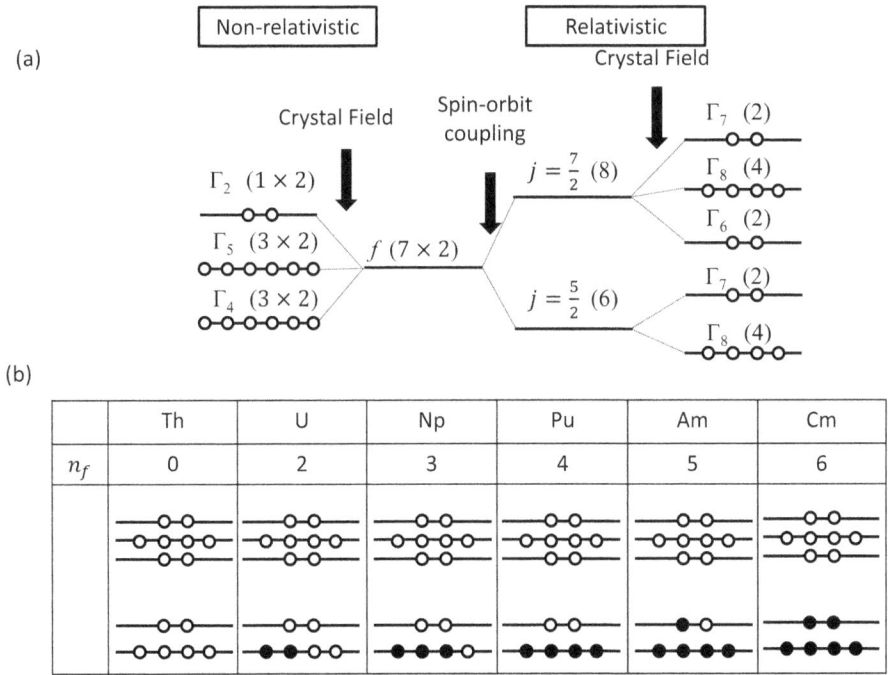

FIGURE 4.2 (a) Splitting of the f-orbitals by SOC and crystal field. The left and right panels show non-relativistic and relativistic cases, respectively. (b) The table of occupation features the f-orbitals for AnO_2 in the relativistic case.

not expected to occur in these materials. In the experiments, no magnetic order was observed in ThO_2 or PuO_2. On the contrary, UO_2, NpO_2, and AmO_2 are considered to have magnetic dipole or multipole orders because of their open shells. In the experiments, a transverse triple-q antiferromagnetic order was observed in UO_2 [8], while a longitudinal triple-q antiferro-multipole order is expected in NpO_2 [9]. AmO_2 is also regarded to exhibit phase transition to a certain ordered state described above at a low temperature [10].

4.2.2 DFT CALCULATION METHODS

The DFT often fails to predict the physical properties of some special groups of materials because the approximation of the exchange-correlation (XC) energy is not suitable for some special materials although it has achieved an accurate prediction on other materials. LDA and GGA, which are exchange-correlation energy approximations used frequently in the DFT, are known to exhibit the self-interaction error in cases of strongly correlated electron systems. To resolve this problem, extended DFT methods, such as DFT+U and hybrid-DFT, are employed. These methods are crucial for the calculations of the electronic structures of AnO_2. In addition, SOC is required for an accurate estimation of the electronic states of 5f-orbitals.

DFT+U has been commonly employed among the methods that incorporate strong correlation effects. In this method, the correlation energy described by Hubbard U is added to the LDA or GGA energy. Its relatively small computational cost is essential for common usage in computational works. The total energy obtained by DFT+U is expressed as follows:

$$E_{DFT+U} = E_{LDA/GGA} + E_U - E_{DC},$$

where $E_{LDA/GGA}$ is the LDA or GGA energy and E_U is the correlation energy obtained by adding Hubbard U. E_{DC} removes the double counting of the correlation energy of $E_{LDA/GGA}$ and E_U. Despite the various models for E_{DC} and E_U that have so far been suggested, the models of Liechtenstein et al. [11] and Dudarev et al. [12] have been mostly employed. However, note that the DFT+U energy may not converge to the true ground state but reach a local minimum [13].

The hybrid-DFT is also commonly used for AnO_2 calculations. In this method, the total electronic energy is obtained by mixing DFT and Hartree–Fock energies. PBE0 [14] and B3LYP [15] have been mostly used among the proposed models of the hybrid-DFT. However, the computational costs of the hybrid-DFT have significantly increased compared to that of DFT+U due to Hartree–Fock energy calculations. Correspondingly, the HSE model [16] was suggested, in which Hartree–Fock energy is evaluated only in the short range of atoms.

In addition to the total energy of materials, various physical properties can also be evaluated using DFTs. For instance, phonon dispersions can be obtained by calculating atomic forces with atom displacements from their stable positions [17]. We can then obtain the specific heat capacity and thermal expansion coefficients using the calculated phonon dispersions [18]. In addition, we can calculate other thermal properties based on the excitation energy of the electronic states, which can also be estimated by the DFTs. In Ref. [19], the specific heat capacity of PuO_2 was estimated by combining the phonon and Schottky heat capacities caused by electronic excitations, both of which were based on DFT calculations.

Ab-initio MD, in which atomic forces are calculated by DFTs in the "on the fly" way, is an effective method of directly calculating physical properties, despite their computational cost being too high for performing large-scale and long-time simulations.

We will next concentrate on the ground-state calculations by DFTs on each oxide fuel material from UO_2 to other actinide oxides. These calculations are crucial because various material properties can be estimated based on the obtained states.

4.2.3 UO_2

UO_2 is one of the main compounds in nuclear fuels and is presently employed in LWRs; thus, a large number of DFT calculations have so far been performed for UO_2 [4]. In most of these calculations, DFT+U [11,12] or hybrid-DFT [16] has been adopted, and the antiferromagnetic insulating electronic states have been predicted. The calculated physical properties, such as lattice constants, bulk modulus, and bandgap, show good agreement with the experimental data [6]. However, some of these calculations were performed using collinear magnetism without SOC [20], although the experimental results showed a non-collinear magnetic order (transverse triple-q antiferromagnetic order, Figure 4.1) in the ground state of UO_2. DFT calculations with SOC for the triple-q antiferromagnetic states were also performed [21,22]. However, single-q and double-q antiferromagnetic states were more stable than the triple-Q state, which disagrees with the experimental results. This discrepancy might be resolved by more improved methods, including many-body correlation effects [23]. However, no successful result showing consistency with all experimental results has yet been made available. The ground-state exploration is still in progress.

4.2.4 PuO_2

PuO_2 is known as a paramagnetic insulator in experiments. SOC is necessary to obtain this insulating state because a closed-shell state cannot be created by four f-electrons in a Pu ion without SOC. When the SOC is neglected, the f-orbitals are split into two 3-degenerate (Γ_4 and Γ_5) and one non-degenerate (Γ_2) orbitals due to the cubic crystal field. In this case, the four f-electrons in the Pu ions cannot fully occupy any orbitals; therefore, the closed-shell state is not reproduced. Considering the spin degrees of freedom, the f-orbitals are split into 6-, 6-, and 2-degenerate orbitals (Figure 4.2a). In the relativistic case, the lowest level of f-orbitals becomes a 4-degenerate state (Γ_8), and the closed shell is created by the four f-electrons, resulting in the paramagnetic insulating state. These insulating states are obtained by DFT+U [21,24,25] and hybrid-DFT [26] with SOC. However, in these calculation methods, the ferromagnetic and antiferromagnetic states are more stable than the paramagnetic states. That is, their calculations fail to reproduce the true ground state. At present, some reports explain why such struggles exist in PuO_2. We believe that more extensive experiments may resolve these problematic issues because non-stoi-chiometry and defects are known to often destroy the magnetic orders. In addition, this issue is also expected to be solved by methods incorporating many-body effects in more essential manners. There is still no consistent computational result with the experimental results in exact theoretical viewpoints [27–29].

4.2.5 OTHER ACTINIDE DIOXIDES

In addition to the primary oxide nuclear fuel material, UO_2, other actinide oxides are important because of their use as MOX fuel materials. They attract fundamental interest in their ground-state exploration of strongly correlated electrons. First, ThO_2 principally does not have any f-electron, and its band-insulating state can be reproduced without SOC or strong correlation effects. Therefore, various benchmark tests have possibly been performed by using standard DFTs to access physical properties. For example, lattice constants agree with experimental data [6]. Authors recently calculated thermal conductivity and its temperature dependence over a wide temperature range [30]. Furthermore, the physical properties up to an extremely high-temperature range close to the melting points have been studied by using MLMD through standard DFTs.

NpO_2 has a longitudinal triple-q octupole order [9]. Such a complex state was reproduced by DFT+U [31]. Although AmO_2 is known to have some ordered state at low temperatures [10], it is not clear what kind of magnetic order emerges. In AmO_2, the ground state has not yet been clarified by the DFT calculations. Meanwhile, the calculations of antiferromagnetic states have been tentatively performed [21,25]. These tentative calculations are still useful, even if the ground states are not yet accessible at present because we can know the limitations of microscopic computations and use their results with uncertainty. CmO_2 has a closed f-shell like ThO_2 and PuO_2. Non-magnetic states are also obtained by DFT+U [21,25]. However, no consensus has been reached as regards the ground state both in the experiments and calculations.

4.3 MD WITH INTERATOMIC POTENTIALS: EMPIRICAL AND MACHINE LEARNING

Keita Kobayashi and Masahiko Machida

Section 4.2 shows that DFT calculations give descriptions of the ground states of actinide oxides depending on the validity of the employed XC functionals. The ground-state description is of great importance as fundamental knowledge on fuel materials, while the detailed information of nuclear fuel materials in high-tempera-ture ranges from its operating to the melting temperature is prerequisite for both the design of reactors and nuclear safety. The MD simulation is a standard method for obtaining the thermal properties of nuclear materials in a high-temperature range. This simulation relies on the quality of the interatomic potential to describe the inter-atomic force in the system of interest. The potential energy surface (PES) directly obtained by the DFT calculation is a preferable choice in obtaining an interatomic potential. However, due to the large computational efforts of the DFT calculations, the system size and the time step of the so-called first-principle MD simulations obtaining the interatomic force in an "on the fly" way are limited to a few hundred atoms and picoseconds. Therefore, empirical interatomic potentials are being widely utilized in MD simulations. MDs with empirical interatomic potentials allow us to conduct simulations with sufficiently large system sizes and long-time steps to obtain the statistical averages on the thermal equilibrium and non-equilibrium properties. In fact, various thermal properties of nuclear materials, such as thermal expansion, thermal conductivity, atomic diffusion, λ-peak anomaly known as Bredig transition [32], and melting behavior of actinide oxides, can be accessed by using MD with empirical interatomic potentials. In this section, we review some features of the empirical potentials employed in UO_2. Finally, we introduce an application of MLMD to ThO_2 [33]. The MLMD is a trial approach for directly extending the limitation of the DFT calculations into MD levels with keeping its accuracy within the DFT levels by learning the DFT calculation results [34, 35]. We show that MLMD is a powerful method for accessing the high-temperature thermophysical behaviors of materials. ThO_2 is a candidate material for demonstrating the usefulness of MLMD among oxide fuel ones because standard DFTs with a relatively low computing cost can accurately capture its ground state [33].

4.3.1 MD Simulations with Empirical Interatomic Potentials

Two types of empirical interatomic potentials are mainly used in classical MD simulations. The first is the shell–core model [36], which explicitly considers the polarization effects of atoms. In this model, an ion consists of a massless charged shell and a massive charged core that represent the electron shells and the atomic nucleus, respectively. The second is the rigid ion model [37], in which ions are simply described as massive point charges. MD simulations with the rigid ion model are faster than those with the shell–core model [36]. The thermal properties of actinide oxides are mainly evaluated with the rigid ion model. This section mainly focuses on the features of the interatomic potentials based on the rigid ion models. The interatomic pairwise potentials of the rigid ion model on actinide oxides are usually designed using relatively simple pairwise functional forms as follows:

$$V_{\text{pair}}\left(r_{ij}\right) = V_C\left(r_{ij}\right) + V_B\left(r_{ij}\right) + V_M\left(r_{ij}\right)$$

$$V_C\left(r_{ij}\right) = q_i q_j e^2 \,/\, 4\pi\epsilon_0 r_{ij}$$

$$V_B\left(r_{ij}\right) = A_{ij}\exp\left(-r_{ij}\,/\,\rho_{ij}\right) - C_{ij}\,/\,r_{ij}^6$$

$$V_M\left(r_{ij}\right) = D_{ij}\left\{\left[1 - \exp\left(\beta_{ij}\left(r_{ij} - r_0\right)\right)\right]^2 - 1\right\},$$

where $V_C(r_{ij})$ is the Coulomb potential with the partial charges q_i of atoms and $V_B(r_{ij})$ is the Buckingham potential [38,39], in which the first term models the repulsive interaction caused by Pauli's exclusion principle for electrons, and the second term represents the van der Waals interaction between the atoms. The combination of the Coulomb and Buckingham potentials has been very often employed as one of the interatomic potentials in oxide materials. The Morse potential on $V_B(r_{ij})$ [40] is also often employed to model the covalent bonds between oxygen and cation atoms. In addition to the above-mentioned simple pairwise potentials, the embedded atom method (EAM) [41] was applied to actinide oxides by Cooper, Rushton, and Grimes (CRG) [42] to include the effects of many-body interactions.

Table 4.1 lists the typical interatomic potentials employed in UO_2. The parameterizations of these empirical interatomic potentials are given by the known experimental data or those obtained by DFT calculations.

Various UO_2 properties, such as lattice constants, elastic properties, and defect formation energies, were investigated by using the empirical interatomic potentials in equilibrium conditions [43]. All interatomic potentials can give reasonable results on the experimental lattice constants and elastic properties. However, this is not necessarily surprising because the empirical potentials are generally fitted to these experimental properties. Moreover, although empirical atomic potentials tend to slightly overestimate the oxygen Frenkel pair (OFP) formation energies compared to the experimental and DFT data, almost all potentials provide OFP energy within the experimental range. In contrast, the interatomic potentials based on the rigid ion model have generally failed to reproduce optical phonon energies due to the lack of polarization effects of the target ions. Meanwhile, the core–shell models can reproduce the optical phonon modes well. A detailed comparison of the equilibrium properties using interatomic potentials is summarized in Ref. [43].

The equilibrium calculations using the interatomic potentials listed in Table 4.1 give similar lattice constants, elastic properties, and defect formation energies. However, MD simulations with different interatomic potentials show significantly different behaviors, as seen in high-temperature thermal properties [43,51,54,55]. For example, the Arima and Walker potentials overestimate the λ-peak

---

TABLE 4.1

Interatomic Potentials Based on the Rigid Ion Models for UO_2

Potential	Coulomb + Buckingham	Coulomb + Buckingham + Morse	Coulomb + Buckingham + Morse + EAM
Author	Arima at al. [44]	Basak et al. [52]	Cooper et al. [42]
	Karakasidis and Lindan [45]	Yamada et al. [53]	
	Lewis and Catlow [46]		
	Morelon et al. [477]		
	Sindzingre and Gillan [48]		
	Tharmalingam [49]		
	Walker and Catlow [50]		
	Potashnikov et al. [51]		

temperature. The Yamada potential predicts the first-order phase transitions between the normal and Bredig phases, whereas the other potentials do not. The Potashnikov and Yakub potentials give reasonable results for the λ-peak temperature, temperature dependence of enthalpy, and thermal expansion of UO_2 in an extremely high-temperature range. However, all the simple pairwise interatomic potentials listed in Table 4.1 largely overestimate the melting point. Thus, the development of common-good interatomic potentials remains a challenging task. To the best of the authors' knowledge, the most reliable empirical interatomic potential currently available for actinide oxides is the CRG potential [42]. The thermal expansion, enthalpy, λ-peak, and melting point computed by the CRG potential show an excellent agreement with the experimental data. Comparing simple pairwise potentials, the success of the CRG potential implies the importance of many-body effects for describing the high-temperature properties in actinide oxides.

4.3.2 MLMD SIMULATIONS FOR ThO_2

As reviewed above, the MD simulation results largely depend on the quality of the empirical interatomic potentials. Depending on the physical quantity of interests and the scope of applications, an empirical interatomic potential must be carefully chosen. An alternative method of calculating the thermal properties of materials is using the first-principle MD simulations. When applying the first-principle MD on ThO_2 [56–59], the enthalpy, thermal expansion, and λ-peak temperature are reasonably estimated. However, the system size and the time step are limited to approximately 300 atoms and a few tens of picoseconds, respectively. The evaluation in terms of phase changes like the melting behaviors is beyond the application of the first-principle MD.

A recent breakthrough in atomic-level MD simulations is the MLMD simulation introduced by Behler and Parrinello [34,35]. In this method, artificial neural networks (ANNs) are utilized to imitate the DFT PES by interpolating a large number of DFT reference data. Although the empirical interatomic potentials consist of simple functional forms based on physical modeling with a few parameters, the interatomic potential using ANNs is not based on any physical modeling, but has a large number of adjustable parameters. The rich flexibility in ANNs enables us to make the PES accuracy comparable to those directly calculated by the DFT.

We briefly introduce herein an MLMD application to ThO_2. In the case of ThO_2, Th cations principally lose all electrons in the outer f-shell, resulting in no f occupation. Therefore, it is not necessary to include the effects of the strongly correlated localized f-electrons in the DFT calculations. We trained the ANNs using the DFT reference data based on the LDA [58,60], Perdew–Burke–Ernzerhof for solids (GGA-PBEsol) [60], and the strongly constrained and appropriately normed (SCAN) [61], meta-GGA XC functionals.

Each ANN's interatomic potential is referred to as ANN–LDA, ANN–PBEsol, and ANN–SCAN. Figure 4.3 shows the phonon dispersion curves using the interatomic potentials of the ANNs. The phonon dispersion curves calculated by both DFTs and ANNs can reproduce the experimental data almost completely. Although the interatomic potential of the ANNs is categorized into the rigid ion model, the ANNs give accurate optical phonon modes in contrast to the empirical interatomic

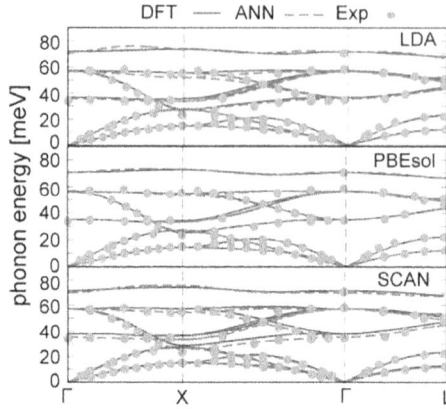

FIGURE 4.3 Phonon dispersion curves of ThO_2 obtained by the DFT (blue line) and the ANN (red line). The results computed by LDA, PBEsol, and SCAN are also shown. The sky-blue dots represent the experimental data [62].

potential based on the rigid model. Table 4.2 summarizes the thermal properties using MLMDs with the ANN interatomic potentials. The maximum system size and the time step in MLMD are 5,184 atoms and 500 ps, respectively, which are far beyond the limit of the first-principle MD simulation. MLMDs provide reasonable results for the coefficient of the linear thermal expansion, λ-peak temperature, and melting point. ANN–SCAN closely reproduces the experimental melting temperature.

We have shown that MLMDs reasonably reproduce various experimental results in high-temperature ranges. Thus, they are considered to be a pathway for determining the detailed thermodynamical behaviors around and above the operating temperature in nuclear reactors. The key issue toward applying MLMD to other actinide oxides is the generation of the DFT reference data based on the proper XC functionals. Thus, it will be crucial to develop XC functionals that will accurately describe the strongly correlated localized f-electron features, as reviewed in Section 4.2.

TABLE 4.2

The Averaged Coefficient of Linear Thermal Expansion (ACLTE) in the Range from 300 to 1,600 K, the λ-Peak Temperature T_λ, and the Melting Point T_m Computed by MD with ANNs Are Summarized

	ANN–LDA	ANN–PBEsol	ANN–SCAN	Exp.
ACLTE (10^{-6} K^{-1})	9.95	10.65	9.71	9.5 [63], 9.67 [64], 11.07 [65]
T_λ (K)	3,040	2,980	3,200	2,950 [66] 3,090 [67]
T_m (K)	3,450–3,460	3,250–3,260	3,610–3,620	3,651 [65]

4.4 BRIEF INTRODUCTORY REVIEW ON MESOSCALE APPROACHES

In studies on nuclear fuel materials, experimental characterizations and tests are essential in evaluating the fuel material properties and their performances. Empirical correlations are usually extracted from the experimental measurement data. On the contrary, theoretical counterparts should always be validated by experimental observations. On the other hand, if theoretical models are fundamentally correct without any empirical assumptions; hence, they can directly predict various material properties. In this case, the only crucial point to which we should pay attention to is their application limitation. However, it is rather difficult to make a commonly useful model to predict material properties and performances without empirical ideas on nuclear fuels. The reason for this is their high complexities produced by irradiation and high-temperature burn-up conditions. The theoretical predictability of simple modeling remains vulnerable, particularly in spent fuel conditions. This strongly motivated us to study the multiscale modeling of fuel materials [2]. In this section, we aim to introduce microstructural evolutions as important phenomena at the mesoscale level and briefly explain the important mesoscale extensions from the abovementioned microscopic modeling as first-principle calculations and MDs. Thus, in the following, we will introduce three modeling methods and their applications in nuclear fuel materials according to a good review paper [2] on multiscale methods.

4.4.1 KINETIC MONTE CARLO METHOD

First, we should mention a scheme that is an important theoretical tool at the mesoscopic scale. This scheme is called the kinetic Monte Carlo (KMC) method. The core of the model is not deterministic, but statistic. The word "statistic" indicates not a dynamic step-by-step procedure, but certain averaging. For example, the statistical scheme is employed to analyze the transition process among system states [68]. In other words, KMC is a kind of coarsening model of MD, that is, the time scale is extended from the atomic vibrations in MD to atom migrations, and the system size scale is expanded to an atom collection. This method clearly drops the vibrations of atoms and focuses on the system's configuration changes. Therefore, the temporal and spatial scales of the KMC simulation are significantly larger than those of MD.

For example, the time period of the scheme is rather long compared to that of MD, and notable ideas are based on two successive evolutions being completely independent and memoryless. The process is called the Markov process. System evolution is controlled by a conversion rate among a set of known possible states. That is, random sampling is performed among various possible transition paths. The transition process from configurations i to j is only related to the transition rate k_{ij}.

KMC is mainly used to simulate radiation damage in studies on nuclear fuels. For example, to simulate the evolution of defects by solving the equations describing the correspondent physical process, system evolution is controlled by a conversion rate among a set of known possible states.

According to Ref. [2], the commonly used KMC methods are object KMC (OKMC) and event KMC (EKMC), which differ in time scale or step size of the single event. The OKMC can be further refined into Atomic KMC (AKMC) and Lattice KMC (LKMC) according to the difference of the main targets. The former mainly deals with atoms, while the latter focuses on the interaction among atoms. KMC has various applications in the nuclear fuel field. AKMC is widely used to study phase transition problems, such as precipitation and phase separation. OKMC is currently being used to study the effects of annealing or temperature changes on damages. EKMC can simulate the evolution of events on a longer time scale. These methods have been employed in studies of fuel and nuclear materials, including the structural materials exposed to radiation.

4.4.2 PHASE FIELD MODEL

Recently, the phase field (PF) model has been frequently employed in a rich variety of fields from materials science to biology and related fields [74]. Its usefulness of conceptual ideas has been examined, even in the high-energy physics fields from elementary particle physics to cosmology [70]. The popular scheme simulates the microstructure evolution, which is a central phenomenon in the core range of the mesoscale behaviors of nuclear fuels. The idea fills the gap between atomic and meso-scales by integrating relevant theoretical concepts on characteristic objects as defects. The PF model can be classified into two types of equations. The first one is called the Cahn–Hilliard (CH) equation, whose evolution target is conservative field variables [72, 73]. The second one is the Allen-Cahn (AC) equation [75, 76], whose PF is a non-conservative one boosted by the time-dependent Ginzburg Landau theory [71].

The first CH mainly treats conservative concentration fields, while the second AC traces non-conservative variables (i.e., order parameters mainly varying on the phase interface). PF models enable us to study the interface diffusion and characterize the microstructure evolution, which are of great significance in predicting the transient behaviors of fuel materials under burn-up conditions.

Through hypothetical modeling, the PF model has been successfully applied to the prediction of complex three-dimensional microstructure evolution dynamics (e.g., solidification, melting, ferroelectricity, ferromagnetic phase transition). In the past 10 years, PF modeling has been widely used to study the microstructure evolution of irradiated nuclear fuels associated with the formation and evolution of bubbles, voids, etc.

4.4.3 RATE THEORY

Nuclear fuels display different evolutional behaviors related to fission products and crystalline defects produced under irradiation conditions. Thus, the calculation of the release rate of fission gases, swelling rate of the cladding cavity, etc. is an important topic in the studies tackling nuclear fuel performance modeling. In such situations, the most crucial modeling issue is the determination of rates like the change rates of reactant or product concentrations in chemical reactions [69]. This means that the rate theory is originated from the chemical reaction rate theory on the basis. Generally, the chemical reaction rate can be given by microscopic calculations using the molecular orbital calculations of the reactant and product molecules in the liquid and gas phases. The rate theory can then be extended to more complex situations in nuclear fuels by similarly giving the rate constants based on the microscopic scheme and other calculation schemes. The rate theory is so mature that the kinetic theory of gases (collision theory) and quantum mechanics (transition state theory) are used to give the rate constants in complex reaction behaviors. In these rate constant determinations, one of the most crucial points is understanding their process evolutions in detail. Nowadays, advanced high-precision microscopic calculations can also be applied to obtain the constants.

In nuclear fuels, the advancements of the rate theory depend on understanding pieces of the reaction processes. In addition, the use of a high-precision microscopic method like the first-principle calculations and MD can further push the advancements.

4.5 HIGH-TEMPERATURE THERMODYNAMICAL PROPERTIES

Masahiko Machida, Keita Kobayashi and Hiroki Nakamura

A nuclear fuel material presently employed in LWRs i.e., UO_2 has been intensively examined using various experimental measurement techniques. MOX fuels, including Pu and other actinides, have also been slightly elaborated by these experimental approaches compared to the rich piled up data in UO_2. However, MOXs with high Pu content (approximately 25–30 at.%) are expected as the reference fuel for Generation IV sodium-cooled fast reactors. The usage of MOXs with a high Pu is indispensable for future spent fuel management through fuel recycling. Thus, the target of the present nuclear fuel research is the assessment of the fuel performance and safety of MOXs together with UO_2 as the basis of actinide oxides. In addition, MOXs with relatively light Pu content have actually been consumed in LWRs. These recent situations have strongly motivated to accurately assess the fuel properties of MOXs and UO_2 and other actinide oxides. Generally, such assessments have been mainly performed by experimental approaches in the early stages of fuel material studies. However, handling Pu and its compounds in experimental schemes remains to be difficult because of its high radiotoxicity [77–80]. Moreover, for other minor actinide elements, the research conditions are almost the same as those for Pu; thus, computational approaches, such as DFT calculations and MD simulations, are expected as possible alternative ways of obtaining various key properties. Experimental schemes further become difficult if the crucial properties are those in high temperatures, including extremely high temperatures close to the melting temperature. Therefore, in this section, we will review the high-temperature behaviors of UO_2, PuO_2, MOXs, and others mainly by using computational approaches.

4.5.1 HEAT CAPACITY

The material properties of UO_2 have been repeatedly reported in various literature. Among them, the heat capacity C_p has been intensively studied by several experimental measurements [81–87]. However, most of these measurements were performed at relatively low temperatures ($T < 2,000$ K), because their targets are those in the operating temperature range of UO_2 in LWRs.

At an extremely high-temperature range close to the melting temperature (T_m), some reports have suggested a peak of heat capacity around $T = 0.8\ T_m$ [84–88]. The peak behavior strongly motivated not only experimentalists but also theorists using advanced computational approaches to study the phenomenon in more detail. This intriguing peak is regarded as a Type II transition in the Ginzburg Landau theory on phase transitions. Historically, it was called the Bredig transition or the λ-transition because it was first discussed by Dworkin and Bredig [32]. However, some groups did not find the peak of the heat capacity due to the complicated measurements at such a high-temperature range [89,90]. This situation has been observed in ThO_2 [67,87] and PuO_2 [91–94]. Moreover, there has been a relatively large scattering in the experimental data C_p in such a rather high-temperature range. According to Ref. [95], the Bredig transition has never been included in any analytical expression for the heat capacity of UO_2 because of the large dispersion in data. The most of proposed available correlations are still monotonically temperature increasing functions.

Pavlov et al. [87] recently suggested a model for the heat capacity of UO_2 validated by their laser flash measurements. It considers a smooth peak related to the Bredig transition. On the contrary, empirical potential calculations (i.e., classical MD [96–99] and molecular Monte Carlo [100] simulations) found the occurrence of the Bredig transition at $T = 0.8\ T_m$ in UO_2. To pursue further accuracy, Cheik Njifon [101], and Yun et al. [102] computed the heat capacity of UO_2 from the first-principle calculations and showed calculation results that were in good agreement with the experimental data. However, no first-principle calculations have been performed at such a higher temperature range, in which Bredig transition was expected to occur. The reason for this is three-folds. First is the extremely high computational costs of the first-principle DFT calculations. Anomalous behaviors due to phase transitions are well known to generally require long-time and large-size simulations to confirm the phase transition origin. Second, the heat capacity is principally a temperature derivative of free energy, which basically requires time-consuming energy calculations in a tiny step of the temperature increment. Third is that UO_2 has highly correlated electronic structures, whose computation is beyond simple standard DFTs. Oxygen movements also occur largely close to the Bredig transition, making the DFT calculations on structures with large deformations not simple.

We propose MLMD as a promising methodology for partially overcoming the abovementioned difficulties. This technique is assigned as a multiscale approach directly extending the scale from the DFT level to the classical MD one. The Bredig transition is a sufficiently-accessible key phenomenon for MLMD. This accessibility can be easily attained because the MLMD can sufficiently reach the size and time scales of a classical MD. Therefore, we performed MLMD on ThO_2 [33], in which the difficulties associated with the strongly correlated effects on f-orbitals are mostly mitigated by the absence of f-orbital occupations in the ground state.

Figure 4.4 shows the molar specific heat capacity computed by MLMD simulations with the ANN potentials based on LDA [58], PBEsol [60], and SCAN [61] XC functionals. The λ-peak positions of the specific heat capacity obtained by the ANN–LDA, ANN–PBEsol, and ANN–SCAN were 3,040, 2,980, and 3,200 K, respectively,

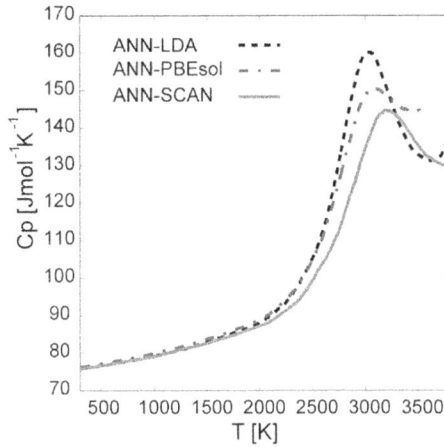

FIGURE 4.4 Temperature dependence of the molar specific heat capacity obtained by the MLMD simulations. The black, blue, and red lines depict the results obtained by MLMD with ANN–LDA, ANN–PBEsol, and ANN–SCAN, respectively.

which were roughly in good agreement with the experimental results of 2,950 K [66] and 3,090 K [67] reported as the Bredig transition temperature.

Reference [95] showed that the properties of the PuO_2 and MOX fuels are poorly known above 2,000 K in experimental measurements. As for PuO_2, the experimental studies available in the literature are only for the heat capacity at low temperatures [92,93,103–105]. In an extremely high-temperature range, Ogard [91] performed his measurement experiment on PuO_2. Subsequently, some authors fitted laws to his data [92,93], often extrapolating their laws to higher temperatures. However, note that the Bredig transition has never been experimentally measured in PuO_2.

In the case of MOX fuels, $U_{1-y}Pu_yO_2$, the available experimental data are also very scarce [91,106–109]. Only the measurements by Leibowitz et al. [107] were performed at a high-temperature range. The most recent measurements were done by Kandan et al. [108]. They performed measurements up to 1,800 K over a range of the Pu content y, such as $0.21 \leq y \leq 0.65$. Gibby et al. established a correlation for the MOX heat capacity by fitting a few available experimental data [91,106,107]. Nevertheless, due to the lack of data, most correlations related to the heat capacity of MOX fuels were determined by applying the Kopp–Neumann rule, that is, the mixing law involving the heat capacity of UO_2 and PuO_2 is given as follows:

$$C_p\left(U_{1-y}Pu_yO_2\right) = \left(1 - y\right)C_p(UO_2) + yC_p(PuO_2).$$

However, no experimental confirmation was provided as regards this law in MOXs. Therefore, computational approaches are strongly demanded to predict its temperature dependencies with an accuracy beyond the classical MD levels.

In addition, the Bredig transition was never experimentally found in any MOX fuel in addition to its one limit side, PuO_2. Hence, all heat capacity correlations of the PuO_2 and MOX fuels available in the literatures have always employed a monotonically increasing function as a function of the temperature.

However, several literature recently reported the heat capacity of $U_{1-y}Pu_yO_2$ in a large range of the Pu content y by using the classical MD with modern empirical potentials from low to extremely high-temperature [96,98,100]. These studies clarified that those materials clearly showed the Bredig transition like UO_2. The latest literature [95] suggested a correlation, including the Bredig transition at 0.84 T_m in the fuel performance analysis. This direction suggests that the existence of the Bredig transition in these MOXs with PuO_2 must be confirmed by experimental measurements and theoretically accurate approaches beyond the classical MD with empirical potentials.

Finally, we mention more details of the Bredig transition regarded to roughly occur at 0.8 T_m [5,100]. This transition can be observed in several ionic materials. These materials discovered about 200 years ago by Faraday [110,111] are called "superionic" materials because they can conserve their ionic composition at a critical temperature. Superionic materials are usually classified as follows: (i) type I superionic materials that exhibit a first-order solid–solid structural transition with a noticeable effect on direct material properties (e.g., lattice parameter of anion sublattice and pair distribution functions), and (ii) type II superionic materials that show a kink in their derivative properties (e.g., heat capacity at constant pressure and thermal expansion coefficient) at a critical temperature (T_λ). At T_λ, the heat capacity at constant pressure and the thermal expansion coefficient exhibit peaks, which are the signature of a second-order or some higher-order "rounded" thermodynamic transition [112]. In actinide oxides, this transition is called the "Bredig transition" or the "lambda transition" [32,86].

So far, all theoretical studies on MOXs were restricted to MD simulations using different empirical potential forms. Yamada et al. [53] published one of the first studies on MOXs that used the Born–Mayer–Huggins interatomic potential [113,114]. In their works, a Morse potential term [53,115] was added to account for the covalent bonding between anions and cations. The authors also evaluated the lattice parameter, heat capacity, compressibility, oxygen diffusivity, and thermal conductivity of PuO_2 and MOX with 0.2% Pu. This study was then extended by Kurosaki et al. [98] to higher temperatures (i.e., up to 3,000 K). Arima et al. [113,116] expanded from the previous studies by using only a Born–Mayer–Huggins interatomic potential for more extensive Pu content ranges. However, these studies failed to reproduce bulk modulus and thermal conductivity at high temperatures of over 2,000 K. Furthermore, the obtained results differ depending on the employed empirical potentials [97,116,117]. These differences indicate that the results strongly depend on the selected potential forms. The parameters on the potential forms were given by fitting lattice parameters, thermal expansion, and elastic constants at room temperature. Cooper et al. [42] recently proposed a more sophisticated form of potential adding a many-body term to the pair-type contributions through the EAM to account for the many-body interactions in the compounds. The results obtained with the CRG potential correctly reproduced many measured properties of actinide oxides and their MOXs for temperatures up to the melting temperature.

MD simulations determine the atom trajectory by solving Newton's equations of motion of the system. During the MD simulations of solids, atoms slightly move around their equilibrium positions; however, the simulation time, which is limited to a few tens of nanoseconds, is not necessarily long enough to observe the atoms moving from one crystalline site to another. This constitutes an MD limitation on the simulations of MOX $U_{1-y}Pu_yO_2$. The stoichiometric $U_{1-y}Pu_yO_2$ MOX is known to form an ideal solid solution [118,119] that is, cations are randomly distributed in the face cubic-centered sublattice. Exploring the cation distribution partition during the simulation requires the system to reach its real thermodynamic equilibrium states. Therefore, using MD to simulate MOXs might not be a relevant approach because the initial cation repartition is kept fixed during the simulation. Alternatively, the molecular Monte Carlo (MC) [120] simulation approach is expected to be more relevant because it enables the exchange of uranium and plutonium atoms during the simulation. The MC advantage allows to explore the space of the cationic configurations in order to achieve a real thermodynamic equilibrium.

4.5.2 Thermal Conductivity

In nuclear fuel research, the important temperature is given by the temperature at the center and periphery of nuclear fuel pellets in both operating and severe accident conditions. The temperature at the central point is generally much higher than that at the periphery. Then, the temperature gradient toward the periphery is crucial in assessing the fuel performance and tolerance of the resultant reactor. In addition, the fission gas release, microstructure cracks, and other phenomena [121] are well known to greatly affect the gradient.

The created temperature gradient is controlled by the local thermal conductivity of fuel materials. Thus, thermal conductivity plays a crucial role in fuel performance evaluations. A number of studies on thermal conductivity have been performed, with the temperature dependence from room temperature and a lower temperature to the normal operating high-temperature, including an extremely high temperature close to the melting temperature. In fact, the temperature-dependent information of thermal conductivity is critical in establishing safe and normal fuel operations (e.g., it gives margins to the fuel central temperature against the melting one, and its transient variations in operations predict fuel performance degradations).

UO_2 is the most widely used nuclear fuel material in conventional LWRs [122–125]; therefore, its thermal conductivity is intensively investigated. The thermal conductivity of UO_2 is relatively low compared to those of other uranium-based compounds, such as metallic uranium [126], uranium nitride [127], uranium carbide [128], uranium silicide [129–132], and uranium–zirconium alloys [133]. Despite its disadvantages, UO_2 is regarded as a commercially useful material because of its high melting point, high ability of fission gas retention, high burn-up capacity, and relatively less reactivity with water and others in contrast to different fuel materials.

The thermal properties of stoichiometric UO_2 have been extensively and intensively studied by both experimental [134–136] and theoretical [20,137–139] researchers UO_2 is basically an electronic insulator at relatively low temperatures of up to ~1,400 K and behaves as a semiconductor at higher temperatures of ~1,400–3,000 K [140] with the presence of some debates on its origin. Consequently, the thermal conduction at the low temperatures of UO_2 is mainly dominated by lattice vibrations below ~1,400 K, while both lattice vibrations and electronic excitations contribute to the thermal transport above ~1,400–3,000 K.

In contrast to the rich information on UO_2, the electronic and thermal transport properties were not sufficiently characterized for hypo-stoichiometric UO_2 fuel materials, albeit the various techniques developed to measure the oxygen-to-uranium ratio to define deviations from stoichiometry [141]. Phonon and electronic characteristics are known to significantly change with the presence of lattice defects. Oxygen vacancies particularly affect the electronic and thermal transport properties of UO_2. Oxygen vacancies are also known to greatly influence electronic transport behaviors, even at room temperature.

The 2011 Fukushima accidents motivated the development of a new series of nuclear fuels with high thermal conductivity. The accidents also inspired the improvement of UO_2 fuels with accident-tolerant fuel characteristics [142–145].

Several studies on accident-tolerant fuels have been performed as seen in Refs. [146,147] since relatively long ago. Consequently, hypo-stoichiometric fuel, i.e., UO_{2-x} is known to exhibit the above characteristics [148,149], such as higher thermal conductivity and lower fission gas diffusion rates compared to stoichiometric UO_2.

The experimental studies of the hypo-stoichiometric phases in UO_2 fuels are few in number in contrast to the stoichiometric UO_2. [13,87,134,150–157]. Moreover, DFT+U studies in hypo-stoichiometric UO_2 fuels related to the other properties, such as electronic properties, electronic thermal conductivity, phonon dispersion, and lattice thermal conductivity, have not yet been performed because fundamental understanding of this partly remains unsolved (as discussed in Section 4.2), even in the stoichiometric one.

To develop a better understanding of the stoichiometric and hypo-stoichiometric UO_2 fuels, we cite a recent literature studying the effect of O vacancies on the electronic and thermal transport properties of UO_2 [5]. The study performed an atomistic simulation in the framework of the DFT+U formalism to investigate the electronic and lattice contributions to thermal conductivity. The computational results for the stoichiometric UO_2 and the hypo-stoichiometric UO_{2-x} were discussed in context with the available experimental results in Ref. [5].

In the DFT+U study of the electronic and thermal properties of the hypo-stoichiometric phases [5], various oxygen vacancy concentrations were examined in a $2\times2\times2$ UO_2 supercell, with $UO_{1.97}$, $UO_{1.94}$, $UO_{1.87}$, $UO_{1.81}$, and $UO_{1.75}$ hypo-stoichiometric cases having one, two, four, six, and eight oxygen vacancies, respectively. At low temperatures (<900 K), thermal conduction is dissipated via phonon scattering caused by impurities and/or point defects. At intermediate temperatures (1,000–1,300 K), the phonon contribution was largely reduced by their detrimental effects on thermal conduction. In contrast, at considerably high temperatures (>1,300 K), the electronic contribution to thermal conductivity becomes significantly dominant instead of the lattice one. Thus, the dependency of both electronic and phonon thermal conductivities on the oxygen vacancy concentration is significant from the viewpoint of fuel performance improvements. The oxygen concentration reduction closes the bandgap and significantly influences the thermal conductivity. A transition from a wide bandgap to a narrow gap and/or metallic state is observed at the low oxygen deficiency. The thermal conductivity increase is then expected considering the substantial increase in electronic contributions. We would like to point out that computational work is essential for their quantitative assessment.

4.6 DEFECT CHEMISTRY AND RELATED HIGH-TEMPERATURE DYNAMICS

Masahiko Machida and Hiroki Nakamura

UO_2 is widely deployed as a nuclear reactor fuel in conventional LWRs because of its radiation tolerance, high melting point, and chemical stability. Among its several advantages, a particularly important feature is its ability to tolerate significant compositional changes without altering its crystal structures. This feasible character is so crucial that it allows to keep large concentrations of soluble fission products and radiation damages without any detrimental volume changes accompanying the phase transitions.

To understand the excellent features permitting large deviations in non-stoichiometric $UO_{2\pm x}$, its so-called "defect chemistry" is widely investigated as a function of the oxygen partial pressure and temperature. The direction is also significant in relation to fuel performance because uranium and oxygen defects affect not only thermal and mass transports but also fission gas release [158–161].

Uranium has a number of f electrons in its ground state as [Rn] $5f^3 6d^1 7s^2$, which allows several valence electronic states with an almost identical energy. This feature also promotes the stability of rich oxidization and reduction phases.

Guéneau et al. [162] modeled the O/U ratio of $UO_{2\pm x}$ for different oxygen partial pressures and temperatures. The results showed that the O/U is reduced when the temperature is increased due to the high entropy of the gaseous O_2 for a fixed oxygen partial pressure. In contrast, the experiments showed that at 1,073 K, the hyper-stoichiometric UO_{2+x} is accommodated by the oxygen interstitial [163,164]. At lower temperatures, UO_{2+x} is known to phase separate into UO_2 and U_4O_9. The interstitial defects in UO_{2+x} then quit acting as the primary mechanism governing the excess oxygen accommodation [165].

These knowledge motivated a number of atomic-scale simulation studies to investigate the behaviors of point defects and their effects on material properties. The microscopic simulations are very suitable for such assessments. The form of the interatomic forces is one of the most crucial issues in ensuring the accuracy of studies in classical MD, while a suitable choice among various DFT frameworks is strongly demanded to capture an accurate electronic structure around the defects (see Section 4.2).

A number of computational works using either classical MD or DFT calculation were performed in the field of defect chemistry. The DFT turned out to be more suitable for defect enthalpy calculations, while the classical MD is required to obtain the vibration entropy contributions by accessing ability in larger system sizes beyond DFT-reachable ones.

Both computational approaches are widely applied to defects in $UO_{2\pm x}$, as presented in many literature [166–179]. Works involving the DFT reported the prediction of uranium vacancy as the dominant defect in UO_{2+x} [174,175] in contradiction to experimental observation [163]. Thus, the omission of vibration entropy in the defect formation energies was suggested as a possible explanation [172]. Using the calculation scheme compensating for the omission, the defect chemistry of PuO_2 was recently investigated. However, to the best of our knowledge, no work on MOX has used the latest scheme, except for $(Pu, Am)O_{2\pm x}$ [176]. These systematic studies will be the next computational target.

4.7 BRIEF NOTES ON MULTISCALE MODELING OF FUEL MATERIALS

Finally, we will briefly review herein the multiscale simulation employed in the research community of nuclear fuel materials. The scheme is composed of a combination of the abovementioned simulation methods with different peculiar times, length scales, applicable phenomena, and processes.

Most of the behaviors in nuclear fuel materials are basically multiscale; hence, various combination works using these methods are widely performed. We briefly introduce the issues here.

First, we will give a short summary of fuel behaviors and suitable approaches. Second, we will discuss their methodological advantages and disadvantages. The list of phenomena and processes of nuclear fuel materials will be classified into heat transport, mass transport, fission gas behaviors, restructuring, and pellet-cladding interactions. These are the essential features of nuclear fuel materials, but there are other relatively minor phenomena and processes as well. For more details, see Ref. [2].

4.7.1 HEAT TRANSPORT

The heat transport in the nuclear fuel [180,181] of operating reactors is important because it is intimately relevant to fuel performance and safety. On the microscopic scale, heat originates from atom vibrations and electronic excitations. They diffuse inside the solid portions of fuel materials. In the former, the atom vibration transport shows diffusive characters, even in a perfect single crystal, due to the anharmonic terms on the interatomic forces. Its theoretical treatment and computation were established in the DFT and classical MD levels; thus, we can estimate its contribution together with its temperature dependence in the single crystal form. The remaining problem is the accuracy of the DFT and the interatomic potentials in MD.

By contrast, the heat transport contribution resulting from electronic excitations is still elusive compared to the atomic vibration counterpart. It emerges only in a high-temperature range because all oxide fuel materials are insulators with large energy gaps in the room temperature range. The electronic contribution estimation is still now a challenging issue. Its calculation necessitates the prediction of excited states, which is basically beyond the DFT from the theoretical viewpoint, albeit some approximate calculations being possible using Boltzmann transport equations. The development of an accurate estimation scheme in an ab-initio manner demands significant advancements in condensed matter physics.

From the engineering viewpoint, it is essential to pay attention to the complex structural inhomogeneities encountered when estimating the heat transport inside the fuel rod unit. The presence of various types of local defects like vacancies and interstitials, as well as dislocations, significantly impedes the heat transport. Moreover, extended defects like grain boundaries, fission gas bubbles, and void structures, greatly reduce the heat transport. Their evolution largely changes the heat transport features with the history of fuel burning. Their estimations can then be performed through modeling by coupling mesoscale-level simulations with atomic-scale calculations.

4.7.2 MATTER TRANSPORT

Matter transports composed of diffusions and dissolutions of uranium (other actinides) and oxygen [182–193] are significant in fuel performance predictions. In addition to the primary element transports, those of fission products like helium and xenon listed in Ref. [2] are also crucial in the fuel performance analysis [194–200]. All existing theoretical models of the abovementioned fission product releasing suggest that the products eventually migrate to the grain boundaries through the transient states after their generation. Herein, we suggest that the temperature, dissolution enthalpy, and migration energy of the fission product element are important factors influencing the migration behaviors. These factors depend on the existing forms of fission product elements (i.e., types and concentrations). Furthermore, the migration energies are dependent on the stoichiometric composition of the fuel materials. In determining these factors with respect to migrations, atomic microscopic calculations are crucial for accurate predictions when estimating the mesoscale phenomena upscaled from the microscopic level. Microscopic calculations play an essential role as the scheme providing the input parameters to the mesoscopic models. For these issues, there are many references in UO_2, as presented in Refs. [201–214]. In the multiscale modeling viewpoint, mesoscopic approaches like RT and PF demand for accurate parameters as far as possible, and DFT and MD calculations play an essential role in modeling matter diffusions and migrations by giving their input fundamental parameters.

4.7.3 FISSION GAS BEHAVIORS

Among the diffusion behaviors of the abovementioned fission products, the transient process of fission gas elements is significant in nuclear fuel performance and safety. The processing phenomena are composed of the following: generation, diffusion, and redissolving of gas atoms; nucleation, growth, and agglomeration of gas bubbles; pinning of bubbles by dislocations and grain boundaries; and merging of bubbles and gas release at the boundaries. The modeling starts with a microscopic local simulation on the bubble nucleation and growth by MD [215, 216] and integrates into PF simulations [217–225]. These works are successful examples of multiscale modeling on nuclear performance and safety analysis.

4.7.4 RESTRUCTURING

In relation to the mass transport of primary and fission product elements, restructuring in the matter phase of nuclear fuel materials is significant. It contains many aspects associated with structural changes like formation and evolution of defects represented by interstitial clusters, dislocation rings, etc., crystallographic evolution as grain refining, grain growth, grain boundary migration, recrystallization, etc., and formation, growth, migration of vacancies, etc.

The structural changes listed above occur at various length scales, and different simulation methods have been applied toward their modeling descriptions. A number of works using MD simulations have also been performed for the simulation of the defect formation behaviors. These studies target not only fuel materials but also cladding and structural materials under irradiation influences. The significant interesting behaviors are high-energy recoil-induced damage [226], high-energy collision cascade [227], time dependence and primary knock-on atom location dependence of post-irradiation defect distribution [228], fission fragment-induced thermal spikes [229], and fission fragment–induced damage [230].

In addition to MD simulation work, KMC has also been employed in cases of displacement collision cascades, thermal spike phases, earlier evolution, transport of defects, etc. [228,231–234]. Its use has an important advantage on the relatively long-time scale for their behaviors.

As for the mesoscale behaviors involving grains and grain boundaries, the use of PF is significantly appealing for the behaviors of grains, boundaries, precipitates, and recrystallization [235–239].

Moreover, RT is also powerful for the formation, growth, and transport of voids regarded as processes analogous to the bubble behaviors simulated by using RT in addition to PF. For their literature, see [217, 225, 240–250].

Finally, we mention the cracking behaviors of fuel materials known as a typical multiscale issue. The reason for this is that once defects occur in a microscopic range, cracks generally tend to propagate and finally lead to visible crack scale or clear failure damages. The modeling of the related intergranular brittle fracture and the fracture strength in UO_2 was reported by Refs. [251] and [252], respectively. The macroscopic analyses based on the FEM were also applied to crack initiation and propagation, with the effects of the grain boundary cracking on the mechanical strength being reported in Refs. [253–255].

4.7.5 PELLET-CLADDING INTERACTION

In nuclear fuel studies, the issues of pellet-cladding interactions are also important because they are related to the reactor failures that affect the design and safety issues of nuclear fuel materials. Their phenomena basically occur as macroscopic mechanical interactions. The occurrences are originated from the combined behaviors of microscopic and mesoscopic scales in nuclear fuel materials. Among the phenomena, the small-scale behaviors are so local that microscopic calculations have an important role. However, more important points are set up in the macro-scale simulations as FEM analyses [256–263].

4.7.6 IMPACTS AND FUTURE DIRECTIONS OF MULTISCALE APPROACHES

Reference [2], which reviewed multiscale schemes on nuclear fuel materials, presented an overwhelming difficulty still existing for the cyclic conveyance of information (calculation results) among simulation methods with various scales.

Practically, the cyclic conveyance of information is rather difficult, while the unidirectional transfer of information from small to large scale is relatively successful. As presented above, the successful examples are so rich, like the integrated simulations from the DFT and MD simulations to the mesoscale calculations, in which various parameters are transferred from microscale to mesoscale. These successes can be partly attributed to R&D of DFT and MD simulations on nuclear fuel materials. In the present decade, their developments have enabled the delivery of rather accurate information. Here, we would like to point out that a new scheme has very recently emerged as an information-delivering scheme for larger-scale simulations. As mentioned in the section on MD simulations, their potentials can be automatically constructed by machine-learning schemes for the DFT calculations [33]. This establishment of automatic ways in information transfer is also applicable to other information-transferring methods to larger scales. We expect machine-learning schemes to play an important role in the future of multiscale simulations on nuclear fuels.

So far, we have mentioned the unidirectional conveyance of information to a larger scale. Here, we now classify a number of inventive challenges summarized according to Ref. [2]: (i) applying methods of different scales to the simulation of a single physical process or phenomenon and comparing the results; (ii) applying methods of different scales to simulation diverse processes or phenomena that involve the same parameters and evaluating the results; (iii) conveyance (virtually still unidirectional) and integration of information from all scales to a single scale; and (iv) coupling of two scales, that is, bidirectional and cyclic conveyance of information between methods of different scales.

As clearly mentioned in the latest review literature [1, 2], the development of a multiscale simulation is not yet at a mature stage, but in a growing stage in the field of nuclear fuel studies. However, technical innovations should be piled up in improving the design and safety analysis of nuclear fuel materials because their phenomena and processes are principally multiscale ones integrated from a microscopic origin to larger-scale behaviors. In the future, we expect such technical challenges of multiscale simulations to continue and discover efficient and effective schemes to bridge the gap between different scales, processes, and phenomena.

As a methodological innovation, we finally suggest that machine-learning techniques are greatly promising. We have already seen fruitful results in bridging DFT to MD simulations [33]. The scheme keeps the DFT accuracy and extends the simulation size and time to classical MD levels [33]. It is easily possible to apply the technique to MC methods without tracing long-time atomic motions. In addition, machine-learning techniques have been intensively employed in surrogating dynamical simulations of continuous variables with high computational costs [264]. This means that the scheme can replace various mesoscale simulations by surrogation models with significantly reduced computational costs. Such technical innovations may make multiscale simulations more feasible and accessible.

ABBREVIATIONS

AC	Allen-Cahn
ACLTE	Averaged coefficient of linear thermal expansion

AKMC	Atomic kinetic Monte Carlo
ANN	Artificial neural network
CH	Cahn-Hilliard
CRG	Cooper-Rushton-Grimes
DFT	Density functional theory
EAM	Embedded atom method
EKMC	Event kinetic Monte Carlo
FEM	Finite element method
GGA	Generalized gradient approximation
HSE	Heyd–Scuseria–Ernzerhof
KMC	Kinetic Monte Carlo
LDA	Local density approximation
LKMC	Lattice kinetic Monte Carlo
LWR	Light-water reactor
MC	Monte Carlo
MD	Molecular dynamics
MLMD	Machine-learning molecular dynamics
MOX	Mixed oxide
OFP	Oxygen Frenkel pair
OKMC	Object kinetic Monte Carlo
PES	Potential energy surface
PF	Phase field
PKA	Primary knock-on atom
RT	Rate theory
SCAN	Strongly constrained and appropriately normed
SOC	Spin–orbit coupling
XC	Exchange-correlation

REFERENCES

1. Bartel, T.J., et al., *State-of-the-Art Report on Multi-scale Modelling of Nuclear Fuels.* No. NEA-NSC-R-2015-5. 2015. Organisation for Economic Co-Operation and Development.

2. Wen, C., et al., Applying multi-scale simulations to materials research of nuclear fuels: A review. *Materials Reports: Energy*, 2021. **1**(3): p. 100048.

3. Hurley, D.H., et al., Thermal energy transport in oxide nuclear fuel. *Chemical Reviews*, 2022. **122**(3): p. 3711–3762.

4. Wen, X.D., et al., Density functional theory studies of the electronic structure of solid state actinide oxides. *Chemical Reviews*, 2013. **113**(2): p. 1063–1096.

5. Kaloni, T.P., et al., DFT+U approach on the electronic and thermal properties of hypostoichiometric UO_2. *Annals of Nuclear Energy*, 2020. **144**: p. 107511.

6. Roy, L.E., Actinides: Electronic structure of the oxides, in *Encyclopedia of Inorganic and Bioinorganic Chemistry*, 2018. https://onlinelibrary.wiley.com/doi/abs/10.1002/9781119951438.eibc2532

7. Santini, P., et al., Multipolar interactions in f-electron systems: The paradigm of actinide dioxides. *Reviews of Modern Physics*, 2009. **81**(2): p. 807–863.

8. Ikushima, K., et al., First-order phase transition in UO_2: ^{235}U and ^{17}O NMR study. *Physical Review B - Condensed Matter and Materials Physics*, 2001. **63**: p. 1044041.

9. Tokunaga, Y., et al., NMR evidence for triple-q multipole structure in NpO_2. *Physical Review Letters*, 2005. **94**(13): p. 137209.

10. Tokunaga, Y., et al., NMR evidence for the 8.5 K phase transition in americium dioxide. *Journal of the Physical Society of Japan*, 2010. **79**(5): p. 053705.

11. Liechtenstein, A.I., V.I. Anisimov, and J. Zaanen, Density-functional theory and strong interactions: Orbital ordering in Mott-Hubbard insulators. *Physical Review B*, 1995. **52**(8): p. R5467–R5470.

12. Dudarev, S.L., et al., Electron-energy-loss spectra and the structural stability of nickel oxide: An LSDA+U study. *Physical Review B*, 1998. **57**(3): p. 1505–1509.

13. Dorado, B., et al., Stability of oxygen point defects in UO_2 by first-principles DFT+ U calculations: Occupation matrix control and Jahn-Teller distortion. *Physical Review B*, 2010. **82**(3): p. 035114.

14. Perdew, J.P., M. Ernzerhof, and K. Burke, Rationale for mixing exact exchange with density functional approximations. *The Journal of Chemical Physics*, 1996. **105**(22): p. 9982–9985.

15. Kim, K. and K.D. Jordan, Comparison of density functional and MP2 calculations on the water monomer and dimer. *The Journal of Physical Chemistry*, 2002. **98**(40): p. 10089–10094.

16. Heyd, J., G.E. Scuseria, and M. Ernzerhof, Hybrid functionals based on a screened Coulomb potential. *Journal of Chemical Physics*, 2003. **118**: p. 8207–8215.

17. Parlinski, K., Z.Q. Li, and Y. Kawazoe, First-principles determination of the soft mode in cubic ZrO_2. *Physical Review Letters*, 1997. **78**(21): p. 4063–4066.

18. Minamoto, S., et al., Calculations of thermodynamic properties of PuO_2 by the first-principles and lattice vibration. *Journal of Nuclear Materials*, 2009. **385**(1): p. 18–20.

19. Nakamura, H., M. Machida, and M. Kato, First-principles calculation of phonon and Schottky heat capacities of plutonium dioxide. *Journal of the Physical Society of Japan*, 2015. **84**(5): p. 053602.

20. Dorado, B., et al., DFT+U calculations of the ground state and metastable states of uranium dioxide. *Physical Review B*, 2009. **79**(23): p. 235125.

21. Suzuki, M.T., N. Magnani, and P.M. Oppeneer, Microscopic theory of the insulating electronic ground states of the actinide dioxides AnO_2 (An = U, Np, Pu, Am, and Cm). *Physical Review B*, 2013. **88**(19): p. 195146.

22. Pegg, J.T., et al., Magnetic structure of UO_2 and NpO_2 by first-principle methods. *Physical Chemistry Chemical Physics*, 2019. **21**(2): p. 760–771.

23. Amadon, B., First-principles DFT+DMFT calculations of structural properties of actinides: Role of Hund's exchange, spin-orbit coupling, and crystal structure. *Physical Review B*, 2016. **94**: p. 115148.

24. Nakamura, H., M. Machida, and M. Kato, Effects of spin-orbit coupling and strong correlation on the paramagnetic insulating state in plutonium dioxides. *Physical Review B*, 2010. **82**.

25. Pegg, J.T., et al., DFT+ U study of the structures and properties of the actinide dioxides. *Journal of Nuclear Materials*, 2017. **492**: p. 269–278.

26. Nakamura, H. and M. Machida, Hybrid density functional study on plutonium dioxide, in *Proceedings of the International Conference on Strongly Correlated Electron Systems (SCES2013)*. 2014.

27. Yang, Y., B. Wang, and P. Zhang, Electronic and mechanical properties of ordered (Pu, U) O_2 compounds: A density functional theory +U study. *Journal of Nuclear Materials*, 2013. **433**(1): p. 345–350.

28. Pegg, J.T., et al., Hidden magnetic order in plutonium dioxide nuclear fuel. *Physical Chemistry Chemical Physics*, 2018. **20**(32): p. 20943–20951.

29. Pegg, J.T., et al., Noncollinear relativistic DFT + U calculations of actinide dioxide surfaces. *The Journal of Physical Chemistry C*, 2018. **123**(1): p. 356–366.

30. Nakamura, H. and M. Machida, First-principles calculation study on phonon thermal conductivity of thorium and plutonium dioxides: Intrinsic anharmonic phonon-phonon and extrinsic grain-boundary–phonon scattering effects. *Journal of Nuclear Materials*, 2019. **519**: p. 45–51.

31. Suzuki, M.T., N. Magnani, and P.M. Oppeneer, First-principles theory of multipolar order in neptunium dioxide. *Physical Review B*, 2010. **82**(24): p. 241103.

32. Dworkin, A.S. and M.A. Bredig, Diffuse transition and melting in fluorite and antifluorite type of compounds. Heat content of potassium sulfide from 298 to 1260.degree.K. *The Journal of Physical Chemistry*, 1968. **72**: p. 1277–1281.

33. Kobayashi, K., M. Okumura, H. Nakamura, M. Itakura, M. Machida, and W.D.M. Cooper, Machine learning molecular dynamics simulations toward exploration of high temperature properties of nuclear fuel materials: case study of thorium dioxide. *Scientific Reports*, 2022. **12**(1): p. 9808.

34. Behler, J. and M. Parrinello, Generalized neural-network representation of high-dimensional potential-energy surfaces. *Physical Review Letters*, 2007. **98**: p. 146401.

35. Behler, J., Constructing high-dimensional neural network potentials: A tutorial review. *International Journal of Quantum Chemistry*, 2015. **115**(16): p. 1032–1050.

36. Dick, B.G. and A.W. Overhauser, Theory of the dielectric constants of alkali halide crystals. *Physical Review*, 1958. **112**(1): p. 90–103.

37. Adams, D.J. and I.R. McDonald, Rigid-ion models of the interionic potential in the alkali halides. *Journal of Physics C: Solid State Physics*, 1975. **8**(13): p. 2198–2198.

38. Buckingham, R., The classical equation of state of gaseous helium, neon and argon. *Proceedings of the Royal Society A: Mathematical, Physical and Engineering Sciences*, 1938. **168**: p. 264–283.

39. Buckingham, R.A. and J.E. Lennard-Jones, The classical equation of state of gaseous helium, neon and argon. *Proceedings of the Royal Society of London. Series A. Mathematical and Physical Sciences*, 1938. **168**: p. 264–283.

40. Morse, P.M., Diatomic molecules according to the wave mechanics. II. Vibrational levels. *Physical Review*, 1929. **34**(1): p. 57–64.

41. Daw, M.S. and M.I. Baskes, Embedded-atom method: Derivation and application to impurities, surfaces, and other defects in metals. *Physical Review B*, 1984. **29**(12): p. 6443–6453.

42. Cooper, M.W.D., M.J.D. Rushton, and R.W. Grimes, A many-body potential approach to modelling the thermomechanical properties of actinide oxides. *Journal of Physics: Condensed Matter*, 2014. **26**(10): p. 105401.

43. Govers, K., et al., Comparison of interatomic potentials for UO_2. Part I: Static calculations. *Journal of Nuclear Materials*, 2007. **366**: p. 161–177.

44. Arima, T., et al., Evaluation of thermal properties of UO_2 and PuO_2 by equilibrium molecular dynamics simulations from 300 to 2000K. *Journal of Alloys and Compounds*, 2005. **400**(1): p. 43–50.

45. Karakasidis, T.E. and P.J.D. Lindan, A comment on a rigid-ion potential for UO_2. *Journal of Physics: Condensed Matter*, 1994. **6**: p. 2965–2969.

46. Lewis, G.V. and C.R.A. Catlow, Potential models for ionic oxides. *Journal of Physics C: Solid State Physics*, 1985. **18**(6): p. 1149–1161.

47. Morelon, N.-D., et al., A new empirical potential for simulating the formation of defects and their mobility in uranium dioxide. *Philosophical Magazine*, 2003. **83**: p. 1533–1555.

48. Sindzingre, P. and M.J. Gillan, A molecular dynamics study of solid and liquid UO_2. *Journal of Physics C: Solid State Physics*, 1988. **21**: p. 4017–4031.

49. Tharmalingam, K., A theoretical study of the energies of formation of point defects in CaF_2 and UO_2. *The Philosophical Magazine: A Journal of Theoretical Experimental and Applied Physics*, 1971. **23**(181): p. 199–204.

50. Walker, J.R. and C.R.A. Catlow, Structural and dynamic properties of UO_2 at high temperatures. *Journal of Physics C: Solid State Physics*, 1981. **14**: p. L979.

51. Potashnikov, S.I., et al., High-precision molecular dynamics simulation of UO_2-PuO_2: Anion self-diffusion in UO_2. *Journal of Nuclear Materials*, 2013. **433**: p. 215–226.

52. Basak, C.B., A.K. Sengupta, and H.S. Kamath, Classical molecular dynamics simulation of UO_2 to predict thermophysical properties. *Journal of Alloys and Compounds*, 2003. **360**(1): p. 210–216.

53. Yamada, K., et al., Evaluation of thermal properties of mixed oxide fuel by molecular dynamics. *Journal of Alloys and Compounds*, 2000. **307**(1–2): p. 1–9.

54. Govers, K., et al., Comparison of interatomic potentials for UO_2: Part II: Molecular dynamics simulations. *Journal of Nuclear Materials*, 2008. **376**: p. 66–77.

55. Potashnikov, S.I., et al., High-precision molecular dynamics simulation of UO_2-PuO_2: Pair potentials comparison in UO_2. *Journal of Nuclear Materials*, 2011. **419**: p. 217–225.

56. Szpunar, B., J.A. Szpunar, and K.-S. Sim, Theoretical investigation of structural and thermo-mechanical properties of thoria. *Journal of Physics and Chemistry of Solids*, 2016. **90**: p. 114–120.

57. Szpunar, B. and J.A. Szpunar, Theoretical investigation of structural and thermomechanical properties of thoria up to 3300 K temperature. *Solid State Sciences*, 2014. **36**: p. 35–40.

58. Ceperley, D.M. and B.J. Alder, Ground state of the electron gas by a stochastic method. *Physical Review Letters*, 1980. **45**(7): p. 566–569.

59. Nakamura, H. and M. Machida, High-temperature properties of thorium dioxide: A first-principles molecular dynamics study. *Journal of Nuclear Materials*, 2016. **478**: p. 56–60.

60. Csonka, G.I., et al., Assessing the performance of recent density functionals for bulk solids. *Physical Review B*, 2009. **79**(15): p. 155107.

61. Sun, J., A. Ruzsinszky, and J.P. Perdew, Strongly constrained and appropriately normed semilocal density functional. *Physical Review Letters*, 2015. **115**(3): p. 036402.

62. Clausen, K.N., et al., Inelastic neutron scattering investigation of the lattice dynamics of ThO_2 and CeO_2. *Journal of the Chemical Society, Faraday Transactions*, 1987. **83**: p. 1109–1112.

63. Momin, A., E.B. Mirza, and M.D. Mathews, High temperature X-ray diffractometric studies on the lattice thermal expansion behaviour of UO_2, ThO_2 and (U0.2Th0.8)O_2 doped with fission product oxides. *Journal of Nuclear Materials*, 1991. **185**: p. 308–310.

64. Rodriguez, P.J. and C.V. Sundaram, Nuclear and materials aspects of the thorium fuel cycle. *Journal of Nuclear Materials*, 1981. **100**: p. 227–249.

65. International Atomic Energy Agency. *Thermophysical Properties Database of Materials for Light Water Reactors and Heavy Water Reactors*. 2006. International Atomic Energy Agency, Vienna.

66. Fischer, D.F., J.K. Fink, and L. Leibowitz, Enthalpy of thorium dioxide to 3400 K. *Journal of Nuclear Materials*, 1981. **102**: p. 220–222.

67. Ronchi, C. and J.P. Hiernaut, Experimental measurement of pre-melting and melting of thorium dioxide. *Journal of Alloys and Compounds*, 1996. **240**: p. 179–185.

68. Xu, H., et al., Self-evolving atomistic kinetic Monte Carlo simulations of defects in materials. *Computational Materials Science*, 2015. **100**: p. 135–143.

69. Xu, Z., *Introduction Theory of Chemical Reaction Rate*. 1987. Jiangsu Science and Technology Press, Nanjing.

70. Volovik, G.E., Superfluid analogies of cosmological phenomena. *Physics Reports*, 2001. **351**(4): p. 195–348.

71. Landau, L.D. and I.M. Khalatikov, *The Selected Works of LD Landau*. 1963. Pregamon, Oxford.

72. Cahn, J.W. and J.E. Hilliard, Free energy of a nonuniform system. I. Interfacial free energy. *The Journal of Chemical Physics*, 1958. **28**(2): p. 258–267.
73. Cahn, J.W., On spinodal decomposition. *Acta Metallurgica*, 1961. **9**(9): p. 795–801.
74. Biner, S.B., An overview of the phase-field method and its formalisms, in *Programming Phase-Field Modeling*, S.B. Biner, Editor. 2017. Springer International Publishing, Cham. p. 1–7.
75. Allen, S.M. and J.W. Cahn, A correction to the ground state of FCC binary ordered alloys with first and second neighbor pairwise interactions. *Scripta Metallurgica*, 1973. **7**(12): p. 1261–1264.
76. Allen, S.M. and J.W. Cahn, Ground state structures in ordered binary alloys with second neighbor interactions. *Acta Metallurgica*, 1972. **20**(3): p. 423–433.
77. Truphémus, T., et al., Structural studies of the phase separation in the UO_2–PuO_2–Pu_2O_3 ternary system. *Journal of Nuclear Materials*, 2013. **432**(1–3): p. 378–387.
78. Truphémus, T., et al., Phase equilibria in the uranium-plutonium-oxygen ternary phase diagram at $(U_{0.55}, Pu_{0.45}) O_{2-x}$ and $(U_{0.45}, Pu_{0.55}) O_{2-x}$. *Procedia Chemistry*, 2012. **7**: p. 521–527.
79. Belin, R.C., et al., In situ high temperature X-Ray diffraction study of the phase equilibria in the UO_2–PuO_2–Pu_2O_3 system. *Journal of Nuclear Materials*, 2015. **465**: p. 407–417.
80. Strach, M., et al., Melting behavior of mixed U–Pu oxides under oxidizing conditions. *Nuclear Instruments and Methods in Physics Research Section B: Beam Interactions with Materials and Atoms*, 2016. **374**: p. 125–128.
81. Groenvold, F., et al., Thermodynamics of the UO_{2+x} phase. I. Heat capacities of $UO_{2.017}$ and $UO_{2.254}$ from 300 to 1000 K and electronic contributions. *The Journal of Chemical Thermodynamics*. 1970. **2**: p. 665–679.
82. Huntzicker, J.J. and E.F. Westrum Jr, The magnetic transition, heat capacity, and thermodynamic properties of uranium dioxide from 5 to 350 K. *The Journal of Chemical Thermodynamics*, 1971. **3**(1): p. 61–76.
83. Engel, T.K., The heat capacities of Al_2O_3, UO_2, and PuO_2 from 300 to 1100 °K. *Journal of Nuclear Materials*, 1969. **31**(2): p. 211–214.
84. Hutchings, M.T., High-temperature studies of UO_2 and ThO_2 using neutron scattering techniques. *Journal of the Chemical Society, Faraday Transactions 2: Molecular and Chemical Physics*, 1987. **83**(7): p. 1083–1103.
85. Hiernaut, J.P., G.J. Hyland, and C. Ronchi, Premelting transition in uranium dioxide. *International Journal of Thermophysics*, 1993. **14**(2): p. 259–283.
86. Ralph, J. and G.J. Hyland, Empirical confirmation of a Bredig transition in UO_2. *Journal of Nuclear Materials*, 1985. **132**: p. 76–79.
87. Pavlov, T., et al., High temperature measurements and condensed matter analysis of the thermo-physical properties of ThO_2. *Scientific Reports*, 2018. **8**(1): p. 1–14.
88. Ronchi, C. and G.J. Hyland, Analysis of recent measurements of the heat capacity of uranium dioxide. *Journal of Alloys and Compounds*, 1994. **213**: p. 159–168.
89. Affortit, C., Chaleur spécifique de UN, UC et UO_2. *High Temperatures – High Pressures*, 1969. **1**: p. 27–33.
90. Affortit, C. and J.P. Marcon, Specific heat of uranium and plutonium oxides at high temperature. *Revue Internationale Des Hautes Températures Et Des Refractaires*, 1970. **7**(3): p. 236–&.
91. Ogard, A. High-temperature heat content of plutonium dioxide, in *The 4th International Conference on Plutonium and Other Actinides*. 1970. Santa Fe, NM.
92. Kruger, O.L. and H. Savage, Heat capacity and thermodynamic properties of plutonium dioxide. *The Journal of Chemical Physics*, 1968. **49**(10): p. 4540–4544.

93. Oetting, F.L., The chemical thermodynamics of nuclear materials. VII. The high-temperature enthalpy of plutonium dioxide. *Journal of Nuclear Materials*, 1982. **105**(2–3): p. 257–261.

94. Fink, J.K., Enthalpy and heat capacity of the actinide oxides. *International Journal of Thermophysics*, 1982. **3**(2): p. 165–200.

95. Bathellier, D., et al., A new heat capacity law for UO_2, PuO_2 and (U, Pu) O_2 derived from molecular dynamics simulations and useable in fuel performance codes. *Journal of Nuclear Materials*, 2021. **549**: p. 152877.

96. Cooper, M.W.D., et al., Thermophysical properties and oxygen transport in the (U_x, Pu_{1-x}) O_2 lattice. *Journal of Nuclear Materials*, 2015. **461**: p. 206–214.

97. Potashnikov, S.I., et al., High-precision molecular dynamics simulation of UO_2-PuO_2: superionic transition in uranium dioxide. arXiv preprint arXiv:1102.1553, 2011.

98. Kurosaki, K., et al., Molecular dynamics study of mixed oxide fuel. *Journal of Nuclear Materials*, 2001. **294**(1–2): p. 160–167.

99. Yakub, E., C. Ronchi, and D. Staicu, Molecular dynamics simulation of premelting and melting phase transitions in stoichiometric uranium dioxide. *The Journal of Chemical Physics*, 2007. **127**(9): p. 094508.

100. Takoukam-Takoundjou, C., E. Bourasseau, and V. Lachet, Study of thermodynamic properties of $U_{1-y}Pu_yO_2$ MOX fuel using classical molecular Monte Carlo simulations. *Journal of Nuclear Materials*, 2020. **534**: p. 152125.

101. Njifon, I.C., et al., Electronic structure investigation of the bulk properties of uranium–plutonium mixed oxides (U, Pu)O_2. *Inorganic Chemistry*, 2018. **57**(17): p. 10974–10983.

102. Yun, Y., D. Legut, and P.M. Oppeneer, Phonon spectrum, thermal expansion and heat capacity of UO_2 from first-principles. *Journal of Nuclear Materials*, 2012. **426**(1): p. 109–114.

103. Sandenaw, T.A., Heat capacity of plutonium dioxide below 325°K. *Journal of Nuclear Materials*, 1963. **10**(3): p. 165–172.

104. Flotow, H.E., et al., Heat capacity of $^{242}PuO_2$ from 12 to 350°K and of $^{244}PuO_2$ from 4 to 25°K. Entropy, enthalpy, and Gibbs energy of formation of PuO_2 at 298.15°K. *The Journal of Chemical Physics*, 1976. **65**(3): p. 1124–1129.

105. Kato, M., et al., Thermal expansion measurement and heat capacity evaluation of hypostoichiometric $PuO_{2.00}$. *Journal of Nuclear Materials*, 2014. **451**(1–3): p. 78–81.

106. Gibby, R.L., *Enthalpy and heat capacity of $U_{75}Pu_{25}O_{2-x}$ (25 to 1490°C)*. 1973: United States. p. Medium: ED; Size: Pages: 36.

107. Leibowitz, L., D.F. Fischer, and M.G. Chasanov, Enthalpy of uranium-plutonium oxides: ($U_{.8}Pu_{.2})O_{1.97}$ from 2350 to 3000 K. *Journal of Nuclear Materials*, 1972. **42**: p. 113–116.

108. Kandan, R., et al., Calorimetric measurements on uranium–plutonium mixed oxides. *Journal of Nuclear Materials*, 2004. **324**: p. 215–219.

109. Kandan, R., R. Babu, K. Nagarajan, and P. R. Vasudeva Rao, Calorimetric measurements on plutonium rich (U,Pu)O_2 solid solutions. *Thermochimica Acta*, 2008. **472**(1): p. 46–49.

110. Chadwick, A.V., Transport in defective ionic materials: from bulk to nanocrystals. *Physica Status Solidi (a)*, 2007. **204**(3): p. 631–641.

111. Faraday, M., *Faraday's Diary*. 1932. G. Bell and Sons, Ltd.

112. Annamareddy, A. and J. Eapen, Mobility propagation and dynamic facilitation in superionic conductors. *The Journal of Chemical Physics*, 2015. **143**(19): p. 194502.

113. Arima, T., et al., Evaluation of thermal conductivity of hypostoichiometric (U, Pu)O_{2-x} solid solution by molecular dynamics simulation at temperatures up to 2000K. *Journal of Alloys and Compounds*, 2006. **415**(1): p. 43–50.

114. Ma, J., et al., Molecular dynamical study of physical properties of $(U_{0.75}Pu_{0.25})O_{2-x}$. *Journal of Nuclear Materials*, 2014. **452**(1): p. 230–234.

115. Basak, C.B. and A.S. Kolokol, A novel pseudo-ion approach in classical MD simulation: A case Study on $(U_{0.8}Pu_{0.2})O_2$ mixed oxide. *Journal of the American Ceramic Society*, 2012. **95**: p. 1435–1439.

116. Arima, T., et al., Equilibrium and nonequilibrium molecular dynamics simulations of heat conduction in uranium oxide and mixed uranium–plutonium oxide. *Journal of Nuclear Materials*, 2008. **376**(2): p. 139–145.

117. Balboa, H., et al., Assessment of empirical potential for MOX nuclear fuels and thermo-mechanical properties. *Journal of Nuclear Materials*, 2017. **495**: p. 67–77.

118. Martin, P., et al., XAS study of $(U_{1-y}Pu_y)$ O_2 solid solutions. *Journal of Alloys and Compounds*, 2007. **444**: p. 410–414.

119. Vigier, J.-F., et al., Structural Investigation of $(U_{0.7}Pu_{0.3})O_{2-x}$ mixed oxides. *Inorganic Chemistry*, 2015. **54**(11): p. 5358–5365.

120. Metropolis, N. and S. Ulam, The Monte Carlo method. *Journal of the American Statistical Association*, 1949. **44**(247): p. 335–341.

121. Bagger, C., M. Mogensen, and C.T. Walker, Temperature measurements in high burnup UO_2 nuclear fuel: Implications for thermal conductivity, grain growth and gas release. *Journal of Nuclear Materials*, 1994. **211**(1): p. 11–29.

122. Burns, P.C., R.C. Ewing, and A. Navrotsky, Nuclear fuel in a reactor accident. *Science*, 2012. **335**(6073): p. 1184–8.

123. Zinkle, S.J. and G.S. Was, Materials challenges in nuclear energy. *Acta Materialia*, 2013. **61**(3): p. 735–758.

124. Zinkle, S.J., et al., Accident tolerant fuels for LWRs: A perspective. *Journal of Nuclear Materials*, 2014. **448**(1–3): p. 374–379.

125. Terrani, K.A., Accident tolerant fuel cladding development: Promise, status, and challenges. *Journal of Nuclear Materials*, 2018. **501**: p. 13–30.

126. Kaity, S., et al., Microstructural and thermophysical properties of U 6wt.%Zr alloy for fast reactor application. *Journal of Nuclear Materials*, 2012. **427**: p. 1–11.

127. Ross, S.B., M.S. El-Genk, and R.B. Matthews, Thermal conductivity correlation for uranium nitride fuel between 10 and 1923 K. *Journal of Nuclear Materials*, 1988. **151**(3): p. 318–326.

128. De Coninck, R., W. Van Lierde, and A. Gijs, Uranium carbide: Thermal diffusivity, thermal conductivity and spectral emissivity at high temperatures. *Journal of Nuclear Materials*, 1975. **57**(1): p. 69–76.

129. Kaloni, T.P. and E. Torres, Thermal and mechanical properties of U_3Si_2: A combined ab-initio and molecular dynamics study. *Journal of Nuclear Materials*, 2020. **533**: p. 152090.

130. White, J.T., et al., Thermophysical properties of U_3Si_2 to 1773K. *Journal of Nuclear Materials*, 2015. **464**: p. 275–280.

131. Antonio, D.J., et al., Thermal and transport properties of U_3Si_2. *Journal of Nuclear Materials*, 2018. **508**: p. 154–158.

132. Jossou, E., M. J. Rahman, D. Oladimeji, B. Beeler, B. Szpunar, and J. Szpunar, Anisotropic thermophysical properties of U_3Si_2 fuel: An atomic scale study. *Journal of Nuclear Materials*, 2019. **521**: p. 1–12.

133. Takahashi, Y., M. Yamawaki, and K. Yamamoto, Thermophysical properties of uranium-zirconium alloys. *Journal of Nuclear Materials*, 1988. **154**(1): p. 141–144.

134. Fink, J.K., Thermophysical properties of uranium dioxide. *Journal of Nuclear Materials*, 2000. **279**(1): p. 1–18.

135. Pang, J.W.L., et al., Phonon density of states and anharmonicity of UO_2. *Physical Review B*, 2014. **89**(11): p. 115132.

136. Saoudi, M., et al., Thermal diffusivity and conductivity of thorium-uranium mixed oxides. *Journal of Nuclear Materials*, 2018. **500**: p. 381–388.

137. Kaur, G., P. Panigrahi, and M.C. Valsakumar, Thermal properties of UO_2 with a nonlocal exchange-correlation pressure correction: a systematic first principles DFT + U study. *Modelling and Simulation in Materials Science and Engineering*, 2013. **21**: p. 065014.

138. Kuzmin, A. and M. Yurkov, Thermal conductivity coefficient UO_2 of theoretical density and regular stoichiometry. *MATEC Web of Conferences*, 2017. **92**: p. 01050.

139. Devey, A. J., First principles calculation of the elastic constants and phonon modes of UO_2 using GGA+U with orbital occupancy control. *Journal of Nuclear Materials*, 2011. **412**(3): p. 301–307.

140. Bates, J.L., C.A. Hinman, and T. Kawada, Electrical conductivity of uranium dioxide. *Journal of the American Ceramic Society*, 1967. **50**(12): p. 652–656.

141. Schaefer, E.A. and J.O. Hibbits, Determination of oxygen-to-uranium ratios in hypo- and hyperstoichiometric uranium dioxide and tungsten--uranium dioxide. *Analytical Chemistry*, 1969. **41**: p. 254–259.

142. Harp, J.M., P.A. Lessing, and R.E. Hoggan, Uranium silicide pellet fabrication by powder metallurgy for accident tolerant fuel evaluation and irradiation. *Journal of Nuclear Materials*, 2015. **466**: p. 728–738.

143. Kim, H.-G., et al., Development status of accident-tolerant fuel for light water reactors in Korea. *Nuclear Engineering and Technology*, 2016. **48**(1): p. 1–15.

144. Che, Y., et al., Modeling of Cr_2O_3-doped UO_2 as a near-term accident tolerant fuel for LWRs using the BISON code. *Nuclear Engineering and Design*, 2018. **337**: p. 271–278.

145. Galloway, J. and C. Unal, Accident-tolerant-fuel performance analysis of APMT steel clad/UO_2 fuel and apmt steel clad/UN-U_3Si_5 fuel concepts. *Nuclear Science and Engineering*, 2016. **182**(4): p. 523–537.

146. Christensen, J.A., et al., Uranium dioxide thermal conductivity. *Transactions of the American Nuclear Society*, 1964. 7: p. 391–392.

147. Hetzler, F.J. and E.L. Zebroski, Thermal conductivity of stoichiometric and hypostoichiometric uranium oxide at high temperatures. *Transactions of the American Nuclear Society*, 1964. 7. https://www.osti.gov/biblio/4688817

148. Yeo, S., et al., Enhanced thermal conductivity of uranium dioxide–silicon carbide composite fuel pellets prepared by Spark Plasma Sintering (SPS). *Journal of Nuclear Materials*, 2013. **433**(1–3): p. 66–73.

149. Braun, J., et al., Chemical compatibility between UO_2 fuel and SiC cladding for LWRs. Application to ATF (Accident-Tolerant Fuels). *Journal of Nuclear Materials*, 2017. **487**: p. 380–395.

150. Harding, J.H. and D.G. Martin, A recommendation for the thermal conductivity of UO_2. *Journal of nuclear materials*, 1989. **166**(3): p. 223–226.

151. Ronchi, C., et al., Thermal conductivity of uranium dioxide up to 2900 K from simultaneous measurement of the heat capacity and thermal diffusivity. *Journal of Applied Physics*, 1999. **85**(2): p. 776–789.

152. Massih, A.R., Electronic transport in pure and doped UO_2. *Journal of Nuclear Materials*, 2017. **497**: p. 166–182.

153. Nichenko, S. and D. Staicu, Molecular dynamics study of the effects of non-stoichiometry and oxygen Frenkel pairs on the thermal conductivity of uranium dioxide. *Journal of Nuclear Materials*, 2013. **433**(1–3): p. 297–304.

154. Wang, B.-T., et al., Thermal conductivity of UO_2 and PuO_2 from first-principles. *Journal of Alloys and Compounds*, 2015. **628**: p. 267–271.

155. Colbert, M., F. Ribeiro, and G. Tréglia, Atomistic study of porosity impact on phonon driven thermal conductivity: Application to uranium dioxide. *Journal of Applied Physics*, 2014. **115**(3): p. 034902.

156. Kim, H.J., M.H. Kim, and M. Kaviany, Lattice thermal conductivity of UO_2 using ab-initio and classical molecular dynamics. *Journal of Applied Physics*, 2014. **115**(12): p. 123510.

157. Hyland, G. J., Thermal conductivity of solid UO_2: Critique and recommendation. *Journal of Nuclear Materials*, 1983. **113**(2): p. 125–132.

158. Catlow, C.R.A., Fission gas diffusion in uranium dioxide. *Proceedings of the Royal Society of London. A. Mathematical and Physical Sciences*, 1978. **364**: p. 473–497.

159. Ball, R. and R.W. Grimes, A comparison of the behaviour of fission gases in $UO_{2\pm x}$ and $U_3O_{8\pm z}$. *Journal of Nuclear Materials*, 1992. **188**: p. 216–221.

160. Andersson, D.A., et al., U and Xe transport in $UO_{2\pm x}$: Density functional theory calculations. *Physical Review B*, 2011. **84**(5): p. 054105.

161. Liu, X.Y., et al., Molecular dynamics simulation of thermal transport in UO_2 containing uranium, oxygen, and fission-product defects. *Physical Review Applied*, 2016. **6**(4): p. 044015.

162. Guéneau, C., et al., Thermodynamic assessment of the uranium–oxygen system. *Journal of Nuclear Materials*, 2002. **304**: p. 161–175.

163. Willis, B.T.M., The defect structure of hyper-stoichiometric uranium dioxide. *Acta Crystallographica Section A*, 1978. **34**(1): p. 88–90.

164. Kang, S.-H., et al., Non-stoichiometry, electrical conductivity and defect structure of hyper-stoichiometric UO_{2+x} at 1000°C. *Journal of Nuclear Materials*, 2000. **277**: p. 339–345.

165. Schaner, B.E., Metallographic determination of the UO_2-U_4O_9 phase diagram. *Journal of Nuclear Materials*, 1960. **2**(2): p. 110–120.

166. Dorado, B. and P. Garcia, First-principles DFT+ U modeling of actinide-based alloys: Application to paramagnetic phases of UO_2 and (U, Pu) mixed oxides. *Physical Review B*, 2013. **87**(19): p. 195139.

167. Dorado, B., et al., First-principles calculations of uranium diffusion in uranium dioxide. *Physical Review B*, 2012. **86**(3): p. 035110.

168. Thompson, A.E. and C. Wolverton, First-principles study of noble gas impurities and defects in UO_2. *Physical Review B*, 2011. **84**(13): p. 134111.

169. Andersson, D.A., et al., Role of di-interstitial clusters in oxygen transport in UO_{2+x} from first principles. *Physical Review B*, 2009. **80**(6): p. 060101.

170. Andersson, D.A., et al., Atomistic modeling of intrinsic and radiation-enhanced fission gas (Xe) diffusion in $UO_{2\pm x}$: Implications for nuclear fuel performance modeling. *Journal of Nuclear Materials*, 2014. **451**(1–3): p. 225–242.

171. Iwasawa, M., et al., First-principles calculation of point defects in uranium dioxide. *Materials Transactions*, 2006. **47**(11): p. 2651–2657.

172. Crocombette, J.-P., First-principles study with charge effects of the incorporation of iodine in UO_2. *Journal of Nuclear Materials*, 2012. **429**(1–3): p. 70–77.

173. Vathonne, E., et al., DFT+ U investigation of charged point defects and clusters in UO_2. *Journal of Physics: Condensed Matter*, 2014. **26**(32): p. 325501.

174. Crocombette, J.-P., et al., Plane-wave pseudopotential study of point defects in uranium dioxide. *Physical Review B*, 2001. **64**(10): p. 104107.

175. Crocombette, J.-P., D. Torumba, and A. Chartier, Charge states of point defects in uranium oxide calculated with a local hybrid functional for correlated electrons. *Physical Review B*, 2011. **83**(18): p. 184107.

176. Neilson, W. D., H. Steele, and S. T. Murphy, Evolving Defect Chemistry of (Pu,Am)$O_{2\pm x}$. *The Journal of Physical Chemistry C*, 2021. **125**(28): p. 15560–15568.

177. Yu, J., R. Devanathan, and W.J. Weber, First-principles study of defects and phase transition in UO_2. *Journal of Physics: Condensed Matter*, 2009. **21**(43): p. 435401.

178. Grimes, R.W. and C.R.A. Catlow, The stability of fission products in uranium dioxide. *Philosophical Transactions of the Royal Society of London. Series A: Physical and Engineering Sciences*, 1997. **335**(1639): p. 609–634.

179. Cooper, M.W.D., S.C. Middleburgh, and R.W. Grimes, Vacancy mediated cation migration in uranium dioxide: The influence of cluster configuration. *Solid State Ionics*, 2014. **266**: p. 68–72.

180. Talip, Z., et al., Thermal diffusion of helium in [238]Pu-doped UO_2. *Journal of Nuclear Materials*, 2014. **445**(1–3): p. 117–127.

181. Anderson, E.E., Radiative heat transfer in molten UO_2 based on the Rosseland diffusion method. *Nuclear Technology*, 1976. **30**(1), p. 65–70.

182. Murphy, S.T., E.E. Jay, and R.W. Grimes, Pipe diffusion at dislocations in UO_2. *Journal of Nuclear Materials*, 2014. **447**(1–3): p. 143–149.

183. Breitung, W., Oxygen self and chemical diffusion coefficients in $UO_{2\pm x}$. *Journal of Nuclear Materials*, 1978. **74**(1): p. 10–18.

184. Sabioni, A.C.S., W.B. Ferraz, and F. Millot, Effect of grain-boundaries on uranium and oxygen diffusion in polycrystalline UO_2. *Journal of nuclear materials*, 2000. **278**(2–3): p. 364–369.

185. Sabioni, A.C.S., W.B. Ferraz, and F. Millot, First study of uranium self-diffusion in UO_2 by SIMS. *Journal of Nuclear Materials*, 1998. **257**: p. 180–184.

186. MacEwan, J.R. and W.H. Stevens, Xenon diffusion in UO_2: some complicating factors. *Journal of Nuclear Materials*, 1964. **11**: p. 77–93.

187. Chroneos, A. and R.V. Vovk, Modeling self-diffusion in UO_2 and ThO_2 by connecting point defect parameters with bulk properties. *Solid State Ionics*, 2015. **274**: p. 1–3.

188. Kim, K.C. and D.R. Olander, Oxygen diffusion in UO_{2-x}. *Journal of Nuclear Materials*, 1981. **102**(1–2): p. 192–199.

189. Matzke, H., Radiation enhanced diffusion in UO_2 and (U, Pu)O_2. *Radiation Effects*, 1983. **75**(1–4): p. 317–325.

190. Matzke, H., On uranium self-diffusion in UO_2 and UO_{2+x}. *Journal of Nuclear Materials*, 1969. **30**(1): p. 26–35.

191. Matzke, H., Diffusion in doped UO_2. *Nuclear Applications*, 1966. **2**(2): p. 131–137.

192. Yajima, S., H. Furuya, and T. Hirai, Lattice and grain-boundary diffusion of uranium in UO_2. *Journal of Nuclear Materials*, 1966. **20**: p. 162–170.

193. Gupta, F., A. Pasturel, and G. Brillant, Diffusion of oxygen in uranium dioxide: A first-principles investigation. *Physical Review B*, 2010. **81**: p. 014110.

194. Miekeley, W. and F.W. Felix, Effect of stoichiometry on diffusion of xenon in UO_2. *Journal of Nuclear Materials*, 1972. **42**(3): p. 297–306.

195. Booth, A.H. and G.T. Rymer, *Determination of the diffusion constant of fission xenon in UO_2 crystals and sintered compacts*. No. AECL-692; CRDC-720, 1958. Atomic Energy of Canada Ltd. Chalk River Project, Chalk River, ON.

196. Une, K., I. Tanabe, and M. Oguma, Effects of additives and the oxygen potential on the fission gas diffusion in UO_2 fuel. *Journal of Nuclear Materials*, 1987. **150**(1): p. 93–99.

197. Schmitz, F. and R. Lindner, Diffusion of heavy elements in nuclear fuels: actinides in UO_2. *Journal of Nuclear Materials*, 1965. **17**(3): p. 259–269.

198. Forsberg, K. and A.R. Massih, Diffusion theory of fission gas migration in irradiated nuclear fuel UO_2. *Journal of Nuclear Materials*, 1985. **135**(2–3): p. 140–148.

199. Cooper, M.W.D., et al., Fission gas diffusion and release for Cr_2O_3-doped UO_2: From the atomic to the engineering scale. *Journal of Nuclear Materials*, 2021. **545**: p. 152590.

200. Galvin, C.O.T., et al., Pipe and grain boundary diffusion of He in UO_2. *Journal of Physics: Condensed Matter*, 2016. **28**(40): p. 405002.

201. Andersson, A.D., et al., *Report on simulation of fission gas and fission product diffusion in UO_2*. No. LA-UR-15-28086. 2016, Los Alamos National Lab. (LANL), Los Alamos, NM.

202. Yun, Y., et al., Atomic diffusion mechanism of Xe in UO_2. *Journal of Nuclear Materials*, 2008. **378**(1): p. 40–44.

203. Yun, Y., O. Eriksson, and P.M. Oppeneer, Theory of He trapping, diffusion, and clustering in UO_2. *Journal of Nuclear Materials*, 2009. **385**: p. 510–516.

204. Liu, X.Y. and D.A. Andersson, Revisiting the diffusion mechanism of helium in UO_2: A DFT+U study. *Journal of Nuclear Materials*, 2018. **498**: p. 373–377.
205. Dorado, B., et al., First-principles calculation and experimental study of oxygen diffusion in uranium dioxide. *Physical Review B*, 2011. **83**(3): p. 035126.
206. Perriot, R.T., et al., Atomistic modeling of out-of-pile xenon diffusion by vacancy clusters in UO_2. *Journal of Nuclear Materials*, 2019. **520**: p. 96–109.
207. Kichigina, N.V., et al. Molecular dynamics simulation of xenon diffusion in UO_2 nanocrystals, in *International Conference "Actual Issues of Mechanical Engineering" 2017 (AIME 2017)*. 2017. Atlantis Press.
208. Cooper, M.W.D., et al. *Milestone Report: The Simulation of Radiation Driven Gas Diffusion in UO_2 at Low Temperature*. No. LA-UR-16-23474. 2016. Los Alamos National Lab.(LANL), Los Alamos, NM.
209. Shekunov, G.S., K.A. Nekrasov, and S.K. Gupta, Molecular dynamics simulation of krypton diffusion in UO_2 nanocrystals. *AIP Conference Proceedings*, 2019. **2174**: p. 020062.
210. Keller, N.V. and K.A. Nekrasov, Molecular dynamics simulation of krypton diffusion in UO2 nanocrystals. *AIP Conference Proceedings*, 2018. 2015(1), p. 020036.
211. Galvin, C.O.T., et al., Oxygen diffusion in Gd-doped mixed oxides. *Journal of Nuclear Materials*, 2018. **498**: p. 300–306.
212. Govers, K., et al., Molecular dynamics simulation of helium and oxygen diffusion in UO2±x. *Journal of Nuclear Materials*, 2009. **395**(1): p. 131–139.
213. Cooper, M.W.D., et al., Modeling oxygen self-diffusion in UO_2 under pressure. *Solid State Ionics*, 2015. **282**: p. 26–30.
214. Sarlis, N.V. and E.S. Skordas, Pressure and temperature dependence of the oxygen self-diffusion activation volume in UO_2 by a thermodynamical model. *Solid State Ionics*, 2016. **290**: p. 121–123.
215. Xiao-Feng, T., L. Chong-Sheng, Z. Zheng-He, and G. Tao, Molecular dynamics simulation of collective behaviour of Xe in UO_2. *Chinese Physics B*, 2010. **19**(5): p. 057102.
216. Huang, M., Molecular dynamic simulation of xenon bubble re-solution in uranium dioxide, M.S. Thesis, 2011. University of Illinois at Urbana-Champaign, USA.
217. Hu, S., et al., Application of the phase-field method in predicting gas bubble microstructure evolution in nuclear fuels. *International Journal of Materials Research*, 2010. **101**: p. 515–522.
218. Hu, S., et al., Phase-field modeling of gas bubbles and thermal conductivity evolution in nuclear fuels. *Journal of Nuclear Materials*, 2009. **392**: p. 292–300.
219. Millett, P.C. and M.R. Tonks, Phase-field simulations of gas density within bubbles in metals under irradiation. *Computational Materials Science*, 2011. **50**: p. 2044–2050.
220. Millett, P.C., A. El-Azab, and D. Wolf, Phase-field simulation of irradiated metals: Part II: Gas bubble kinetics. *Computational Materials Science*, 2011. **50**(3): p. 960–970.
221. Millett, P.C., et al., Phase-field simulation of intergranular bubble growth and percolation in bicrystals. *Journal of Nuclear Materials*, 2012. **425**(1): p. 130–135.
222. Millett, P.C., M.R. Tonks, and S.B. Biner, Mesoscale modeling of intergranular bubble percolation in nuclear fuels. *Journal of Applied Physics*, 2012. **111**: p. 083511.
223. Li, Y., et al., Phase-field simulations of intragranular fission gas bubble evolution in UO2 under post-irradiation thermal annealing. *Nuclear Instruments and Methods in Physics Research. Section B, Beam Interactions with Materials and Atoms*, 2013. **303**: p. 62–67.
224. Hu, S., D. E. Burkes, C. A. Lavender, D. J. Senor, W. Setyawan, and Z. Xu, Formation mechanism of gas bubble superlattice in UMo metal fuels: Phase-field modeling investigation. *Journal of Nuclear Materials*, 2016. **479**: p. 202–215.
225. Rokkam, S., et al., Phase field modeling of void nucleation and growth in irradiated metals. *Modelling and Simulation in Materials Science and Engineering*, 2009. **17**: p. 064002.

226. Tian, X.F., et al., Dynamical simulations of radiation damage induced by 10keV energetic recoils in UO_2. *Nuclear Instruments and Methods in Physics Research Section B: Beam Interactions with Materials and Atoms*, 2011. **269**(15): p. 1771–1776.

227. Zhou, W., et al., Molecular dynamics simulations of high-energy displacement cascades in hcp-Zr. *Journal of Nuclear Materials*, 2018. **508**: p. 540–545.

228. Wang, C., et al., Molecular dynamics study on the irradiance properties of SiC/C interface. *Acta Physica Sinica*, 2014. **63**(15): p. 153402–153402.

229. Huang, M., D. Schwen, and R.S. Averback, Molecular dynamic simulation of fission fragment induced thermal spikes in UO_2: Sputtering and bubble re-solution. *Journal of Nuclear Materials*, 2010. **399**: p. 175–180.

230. Devanathan, R., Molecular dynamics simulation of fission fragment damage in nuclear fuel and surrogate material. *MRS Advances*, 2017. **2**(21): p. 1225–1230.

231. Schwen, D. and R.S. Averback, Intragranular Xe bubble population evolution in UO_2: A first passage Monte Carlo simulation approach. *Journal of Nuclear Materials*, 2010. **402**(2): p. 116–123.

232. Xu, D., et al., Defect microstructural evolution in ion irradiated metallic nanofoils: Kinetic Monte Carlo simulation versus cluster dynamics modeling and in situ transmission electron microscopy experiments. *Applied Physics Letters*, 2012. **101**: p. 101905.

233. De Bellefon, G.M. and B.D. Wirth, Kinetic Monte Carlo (KMC) simulation of fission product silver transport through TRISO fuel particle. *Journal of Nuclear Materials*, 2011. **413**(2): p. 122–131.

234. Gao, Y., et al., Theoretical prediction and atomic kinetic Monte Carlo simulations of void superlattice self-organization under irradiation. *Scientific Reports*, 2018. **8**(1): p. 6629.

235. Millett, P.C., et al., Phase-field simulation of thermal conductivity in porous polycrystalline microstructures. *Journal of Applied Physics*, 2008. **104**: p. 033512.

236. Ahmed, K., et al., Phase field simulation of grain growth in porous uranium dioxide. *Journal of Nuclear Materials*, 2014. **446**(1): p. 90–99.

237. Tonks, M.R., et al., Demonstrating the temperature gradient impact on grain growth in UO_2 using the phase field method. *Materials Research Letters*, 2014. **2**(1): p. 23–28.

238. Mei, Z.-G., et al., Grain growth in U–7Mo alloy: A combined first-principles and phase field study. *Journal of Nuclear Materials*, 2016. **473**: p. 300–308.

239. Liang, L., et al., Mesoscale model for fission-induced recrystallization in U-7Mo alloy. *Computational Materials Science*, 2016. **124**: p. 228–237.

240. Millett, P.C., et al., Void nucleation and growth in irradiated polycrystalline metals: a phase-field model. *Modelling and Simulation in Materials Science and Engineering*, 2009. **17**(6): p. 064003.

241. Li, Y., et al., Phase-field modeling of void evolution and swelling in materials under irradiation. *Science China Physics, Mechanics and Astronomy*, 2011. **54**: p. 856–865.

242. Millett, P.C., et al., Phase-field simulation of irradiated metals: Part I: Void kinetics. *Computational Materials Science*, 2011. **50**(3): p. 949–959.

243. Semenov, A.A. and C.H. Woo, Interfacial energy in phase-field emulation of void nucleation and growth. *Journal of Nuclear Materials*, 2011. **411**(1): p. 144–149.

244. Semenov, A.A. and C.H. Woo, Phase-field modeling of void formation and growth under irradiation. *Acta Materialia*, 2012. **60**(17): p. 6112–6119.

245. Xiao, Z.H., et al., Single void dynamics in phase field modeling. *Journal of Nuclear Materials*, 2013. **439**(1): p. 25–32.

246. Wang, N., et al., Asymptotic and uncertainty analyses of a phase field model for void formation under irradiation. *Computational Materials Science*, 2014. **89**: p. 165–175.

247. El-Azab, A., et al., Diffuse interface modeling of void growth in irradiated materials. Mathematical, thermodynamic and atomistic perspectives. *Current Opinion in Solid State & Materials Science*, 2014. **18**: p. 90–98.

248. Semenov, A.A. and C.H. Woo, Modeling void development in irradiated metals in the phase-field framework. *Journal of Nuclear Materials*, 2014. **454**(1): p. 60–68.
249. Hochrainer, T. and A. El-Azab, A sharp interface model for void growth in irradiated materials. *Philosophical Magazine*, 2015. **95**: p. 948–972.
250. Hu, S. and C.H. Henager, Phase-field modeling of void lattice formation under irradiation. *Journal of Nuclear Materials*, 2009. **394**: p. 155–159.
251. Chakraborty, P., Y. Zhang, and M.R. Tonks, Multi-scale modeling of microstructure dependent intergranular brittle fracture using a quantitative phase-field based method. *Computational Materials Science*, 2016. **113**: p. 38–52.
252. Zhang, Y., et al., Crack tip plasticity in single crystal UO_2: Atomistic simulations. *Journal of Nuclear Materials*, 2012. **430**: p. 96–105.
253. Williamson, R., *Simulating Dynamic Fracture in Oxide Fuel Pellets Using Cohesive Zone Models*. No. INL/CON-08-14787. 2009. Idaho National Lab.(INL), Idaho Falls, ID.
254. Spencer, B.W., et al., *Discrete Modeling of Early-Life Thermal Fracture in Ceramic Nuclear Fuel*. No. INL/CON-14-31355. 2014. Idaho National Lab (INL), Idaho Falls, ID.
255. Huang, H., B.W. Spencer, and J.D. Hales, Discrete element method for simulation of early-life thermal fracturing behavior in ceramic nuclear fuel pellets. *Nuclear Engineering and Design*, 2014. **278**: p. 515–528.
256. Qi, F., et al., Pellet-cladding mechanical interaction analysis of Cr-coated Zircaloy cladding. *Nuclear Engineering and Design*, 2020. **367**: p. 110792.
257. Michel, B., et al., 3D fuel cracking modelling in pellet cladding mechanical interaction. *Engineering Fracture Mechanics*, 2008. **75**(11): p. 3581–3598.
258. Denis, A. and A. Soba, Simulation of pellet-cladding thermomechanical interaction and fission gas release. *Nuclear Engineering and Design*, 2003. **223**(2): p. 211–229.
259. Sercombe, J., R. Masson, and T. Helfer, Stress concentration during pellet cladding interaction: Comparison of closed-form solutions with 2D(r, θ) finite element simulations. *Nuclear Engineering and Design*, 2013. **260**: p. 175–187.
260. Kim, H.C., et al., Development of NUFORM3D module with FRAPCON3.4 for simulation of pellet-cladding mechanical interaction. *Nuclear Engineering and Design*, 2017. **318**: p. 61–71.
261. Michel, B., et al., Simulation of pellet-cladding interaction with the PLEIADES fuel performance software environment. *Nuclear Technology*, 2013. **182**: p. 124–137.
262. Hong, K., J. R. Barber, M. D. Thouless, and W. Lu, Effect of power history on pellet-cladding interaction. *Nuclear Engineering and Design*, 2020. **358**: p. 110439.
263. Brochard, J., et al., Modelling of pellet cladding interaction in PWR fuel, in *Proc. 16th Int. Conf. on 'Structural Mechanics in Reactor Technology' (SMiRT 16), Washington, DC, USA*, 2001: p. 1314.
264. Herman, E., J.A. Stewart, and R. Dingreville, A data-driven surrogate model to rapidly predict microstructure morphology during physical vapor deposition. *Applied Mathematical Modelling*, 2020. **88**: p. 589–603.

5 Fuel Technologies for Irradiation Performance Analysis

CONTENTS

5.1 RELATIONSHIP BETWEEN MODEL AND PROPERTIES

Chapter 3 describes the relational equations to explain the properties using a scientifically descriptive model for application to fuel performance codes. This code can be interpolated and extrapolated in properties depending on the composition. The derived equations are shown as follows:

$$a = 4/\sqrt{3} \cdot \{r_c(1+0.112x) + r_a\}(\text{Å}) \tag{5.1}$$

$$\rho_{th} = \frac{4\bar{M}}{A_v \cdot a^3} \tag{5.2}$$

$$\text{LTE} = \left(C_U \bullet (LTE)_{UO_2} + C_{Pu} \bullet (LTE)_{PuO_2} + C_{Am} \bullet (\text{LTE})_{AmO_2} + C_{Np} \bullet (LTE)_{NpO_2} \right)$$
$$\bullet (1+0.59614 \bullet x) \tag{5.3}$$

$$O/M = 2 + [Oi''] - [Vo^{\cdot\cdot}] = 2 + \left[\left\{ (K_{1/2})_U \cdot P_{O_2}^{\frac{1}{2}} \right\}^{-5} + \left\{ (K_{1/2})_{Pu} \cdot P_{O_2}^{\frac{1}{2}} \right\}^{-5} \right]^{-1/5}$$

$$- \left[\left\{ (K_{-1/2})_U \cdot P_{O_2}^{-\frac{1}{2}} \right\}^{-5} + \left\{ (K_{-1/2})_M \cdot P_{O_2}^{-\frac{1}{2}} \right\}^{-5} + \left\{ (K_{-1/2})_{Pu} \cdot P_{O_2}^{-\frac{1}{2}} \right\}^{-5} \right. \tag{5.4}$$

$$\left. + \left\{ ((K_{-1/4})_U)^{1/2} \cdot P_{O_2}^{-\frac{1}{4}} \right\}^{-5} + \left\{ (2 \cdot K_{-1/3})^{1/3} \cdot P_{O_2}^{-\frac{1}{3}} \right\}^{-5} + \left\{ ((K_{-1/4})_{Pu})^{1/2} \cdot P_{O_2}^{-\frac{1}{4}} \right\}^{-5} + \left\{ C_{Pu}'/2 \right\}^{-5} \right]^{-1/5}$$

DOI: 10.1201/9781003298205-5

$$D^* = D_{Vo}^0 [V_o] \exp\left(-\frac{\Delta H_{Vo}^m}{RT}\right) + 2D_{Oi}^0 [O_i] \exp\left(-\frac{\Delta H_{Oi}^m}{RT}\right) \tag{5.5}$$

$$\tilde{D} = \frac{2 \pm x}{2x} D^* \left(\pm \frac{\partial log P_{O_2}}{\partial \log x}\right), \tag{5.6}$$

$$\sigma_{el} = \sigma_n + \sigma_p = e[e]\mu_n + e[h]\mu_p = \frac{C_{\mu_e}}{T^{\frac{3}{2}}} [e] \exp\left(\frac{E_{\mu_e}}{RT}\right) + \frac{C_{\mu_h}}{T^{\frac{3}{2}}} [h] \exp\left(\frac{E_{\mu_h}}{RT}\right). \tag{5.7}$$

$$V_l = \left\{ \left(C_U \bullet (V_l)_{UO_2} + C_{Pu} \bullet (V_l)_{PuO_2} + C_{Am} \bullet (V_l)_{AmO_2} + C_{Np} \bullet (V_l)_{NpO_2} \right) \right.$$
$$\left. \bullet (1 - 0.7279 \bullet x) \bullet (1 - 1.3172 \bullet P) \right\} \tag{5.8}$$

$$V_t = \left\{ \left(C_U \bullet (V_t)_{UO_2} + C_{Pu} \bullet (V_t)_{PuO_2} + C_{Am} \bullet (V_t)_{AmO_2} + C_{Np} \bullet (V_t)_{NpO_2} \right) \right.$$
$$\left. \bullet (1 - 1.0545 \bullet x) \bullet (1 - 0.8549 \bullet P) \right\} \tag{5.9}$$

$$C_p = C_l + C_d + C_{sch} + C_{exe} \tag{5.10}$$

$$\lambda = \lambda_p + \lambda_e = \frac{1}{A + BT} + \left(\frac{\Delta H_i}{e}\right)^2 \cdot \sigma / (4T) \tag{5.11}$$

The temperature dependence of properties showed the dominant mechanism of thermal conductivity and significant heat changes around 1,500 K. Properties at temperatures lower than 1,500 K were described using a, V_l, V_t, and linear thermal expansion (LTE). These parameters can calculate T_D, γ, and Table 3.7 in Chapter 3 summarizes the mechanical properties. Using Slack's equation, all parameters needed to represent λ_p as functions of Pu content, minor actinide (MA) content, oxygen-to-metal (O/M) ratio, ρ, and temperature have been obtained. $C_l + C_d + C_{sch}$ represents C_p. C_l and C_d could be evaluated from T_D, γ, and LTE, and C_{sch} was obtained from density-functional theory simulation. In this way, we could evaluate the properties at temperatures lower than 1,500 K, and the calculated data correlated well with the experimental data. In this temperature range, $C_l + C_d$ was similar among UO_2, (U, Pu)O_2, and PuO_2. The difference between the materials was caused by C_{sch}, related to 5f electrons. The C_p of PuO_2 is approximately 10 J/mol K higher than that of UO_2, and the MOX data were evaluated using Kopp's law.

In the temperature region higher than 1,500 K, parameters λ and C_p increase sharply, related to the lattice defect concentration. It was considered to form oxygen Frenkel defect and electron–hole pairs as lattice defects. Oxygen potential data were analyzed based on defect equilibrium, and the Brouwer diagram was constructed. The analysis results revealed increased lattice defect concentrations with temperature, and the electronic defect dominated the Frenkel defect. The change in electronic

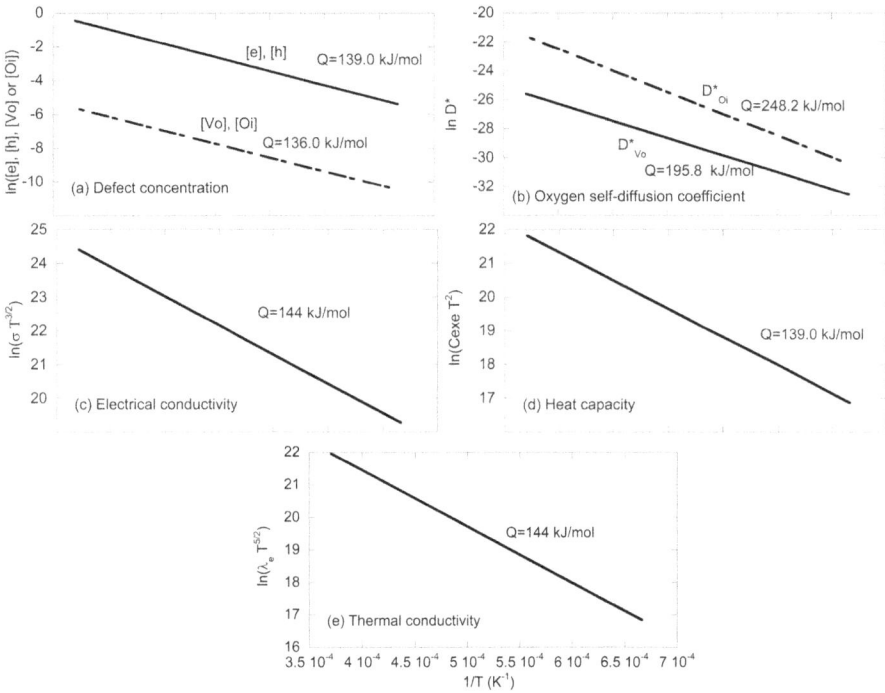

FIGURE 5.1 Relationship between properties and T^{-1}.

defect concentration consistently evaluated λ, C_p, and σ. The Frenkel defect's formation energy was used to evaluate the oxygen diffusion coefficient.

Figure 5.1 shows the properties of $(U_{0.7}Pu_{0.3})O_{2.00}$ against T^{-1}. Activation energies Q for the properties were obtained from a slope in the figure, described by Q/R as follows:

$$Q_e = \Delta H_i/2 = \frac{278}{2} = 139 \text{ kJ/mol} \tag{5.12}$$

$$Q_F = \Delta H_F/2 = \frac{272}{2} = 136 \text{ kJ/mol} \tag{5.13}$$

$$Q_{D_{Oi}^*} = Q_F + \Delta H_{Oi}^m = 136 + 112 \text{ kJ/mol} = 248.2 \text{ kJ/mol} \tag{5.14}$$

$$Q_{D_{Vo}^*} = Q_F + \Delta H_{Vo}^m = 136 + 60 \text{ kJ/mol} = 196 \text{ kJ/mol} \tag{5.15}$$

$$Q_\sigma = Q_e + E_{\mu_e} = 139 + 5 \text{ kJ/mol} = 144 \text{ kJ/mol} \tag{5.16}$$

$$Q_{Cexe} = Q_e = 139 \text{ kJ/mol} \tag{5.17}$$

$$Q_\lambda = Q_e + E_{\mu_e} = 139 + 5 \text{ kJ/mol} = 144 \text{ kJ/mol} \tag{5.18}$$

Q_e and Q_F were 139 and 136 kJ / mol, respectively. Q_F was used to evaluate $Q_{D_{Oi}^*}$ and $Q_{D_{Vo}^*}$, and Q_e represents σ_{el}, C_p, and λ. The analysis results showed that activation energies were consistent among properties. The activation energies obtained in Figure 5.1 corresponded to those obtained in experiments, confirming that the properties described in Chapter 3 are related. The changes in the properties depending on the Pu content, MA content, and O/M ratio can be consistently defined. By applying these formulas to the fuel performance evaluation code, the irradiation behavior of various fuels can be analyzed.

5.2   BREDIG TRANSITION REPRESENTATION

Masato Kato

The specific heat of UO_2 rises sharply and drops at high temperatures, called the Bredig transition. The Bredig transition is a secondary transition with no phase transformation. Such changes in specific heat are also observed in other fluorite compounds, such as CaF_2. The characteristic comparison of CeO_2 and PuO_2 showed they are ionic and electron conductors, respectively [1,2]. Changes in these oxides' specific heat and electrical conductivity at high temperatures were explained by changes in the Frenkel and electronic defects, respectively. Electron conduction appears at high temperatures in actinide oxides due to $5f$ electrons. Therefore, the Bredig transition was evaluated from the change in the electronic defect concentration in this work. The high-temperature terms in heat capacity and thermal conductivity of UO_2 and MOX were represented.

In the current property model (Chapter 3), property changes caused by the Bredig transition have not been presented, vital in property representation. Experimental data at high temperatures were limited, and no data were related to the Bredig transition in MOX properties. Therefore, the equations derived in Chapter 3 should be used in temperature regions of less than the Bredig transition.

The formation energy of e–h was used to evaluate C_p at high temperatures. ΔH_i and ΔS_i of UO_2 were 300 and 85 J/mol K, respectively, in Section 3.3. $[e]$ can be represented by

$$[e] = K_i^{\frac{1}{2}} = \exp(\Delta S_i/2R)\exp(-\Delta H_i/2R)$$

$$= \exp(85/2R)\exp(-300,000/2RT) \tag{5.19}$$

$\Delta H_{\text{Enthalpy}}$ and $C_{P \text{ exe}}$ can be calculated using Eqs. (3.61) and (3.62) due to the increase in $[e]$.

$$\Delta H_{\text{Enthalpy}} = [e]\Delta H_i \tag{5.20}$$

$$C_{P \text{ exe}} = d\Delta H_{\text{Enthalpy}}/dT \tag{5.21}$$

According to Eqs. (5.19) and (5.20), the formation energy of e–h must change at the Bredig transition. It is assumed that ΔH_i and ΔS_i are changed to 100 and 14 J/mol K, respectively, above the Bredig transition temperature. The $[e]$, $\Delta H_{Enthalpy}$, and C_P were described as shown in Figures 5.2–5.4, respectively. Figure 5.4 shows the experimental data reported by Pavlov et al. The calculation results represent the experimental data. The change in $[e]$ at the Bredig transition affects C_P and other properties, such as λ, σ_{el}, $\Delta \bar{G}_{O2}$, and D^*. The λe can be calculated using Eq. (5.22).

$$\lambda_e = \frac{\Delta H_i^2}{T^{2.5}} \cdot D \cdot K_i^{1/2} \cdot \exp\left(-\frac{\mu}{RT}\right) \tag{5.22}$$

It was assumed that D and μ are constant. $K_i^{1/2}$ is obtained by $[e]$ (Figure 5.2) because $[e] = K_i^{\frac{1}{2}} \cdot \lambda$ was represented, as shown in Figure 5.5. The peak related to the Bredig transition of UO_2 was observed at 2,625 K. A better understanding of MOX fuels' Bredig transitions requires experimental data of C_P and σ_{el} up to a melting temperature, but it has never been reported. Here the formation energy of electron–hole pairs

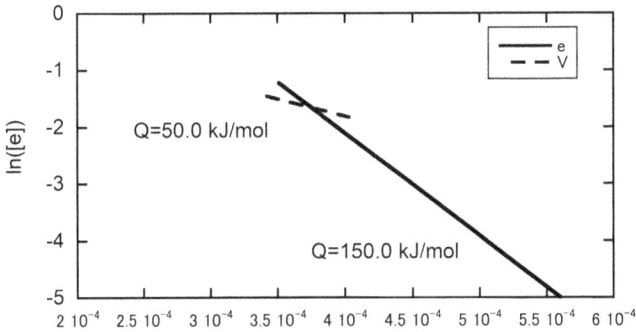

FIGURE 5.2   Relationship between ln([e]) and 1/T in UO_2.

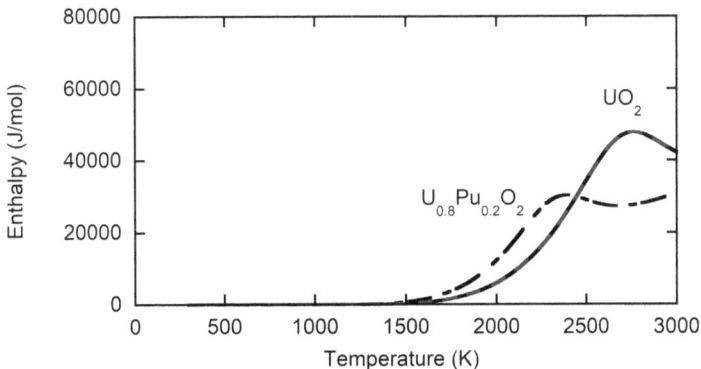

FIGURE 5.3   Enthalpy obtained from the change in [e] with temperature.

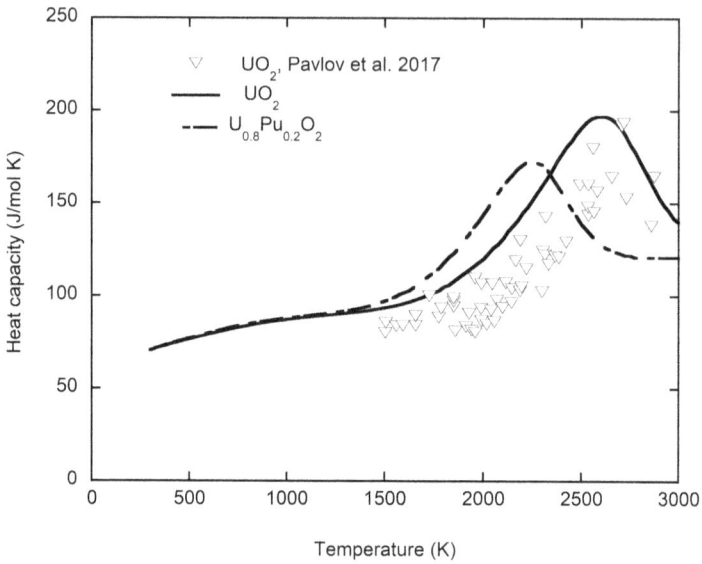

FIGURE 5.4   Heat capacity of UO_2.

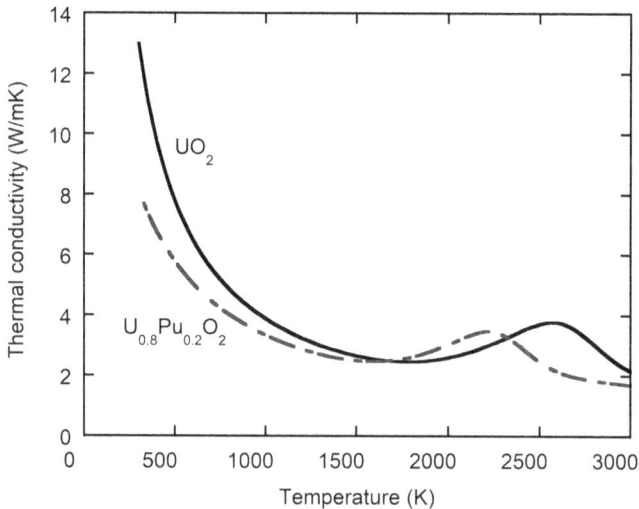

FIGURE 5.5   Thermal conductivity of UO_2 and mixed oxide (MOX) (100% TD) considering the Bredig transition.

in MOX at temperatures higher than the Bredig transition was assumed the same as UO_2. Figures 5.3–5.5 show the calculation results of $(U_{0.8}Pu_{0.2})O_2$. The C_P and λ temperature dependence showed peaks related to the Bredig transition. The Bredig transition temperature was observed at 2,300 K, approximately 300 K lower than UO_2. Experiments and computational science must verify these results.

5.3   TEMPERATURE PROFILE ANALYSIS

Masato Kato, Takayuki Ozawa, Yoshihisa Ikusawa,
Shinji Sasaki, and Koji Maeda

During irradiation, various phenomena, such as restructuring, density change, pore migration, element diffusion, and crack generation, occur in the fuel pellet. Understanding and describing these phenomena are crucial in fuel performance analysis. For example, changes in the microstructure and density affect fuel's thermal conductivity; therefore, knowing when and how fast these changes occur during the irradiation period is vital to know the maximum temperature of the fuel. Figure 5.6 shows the microstructure in the cross section of sintered and irradiated pellets [1,2]. The fresh sintered pellet has a homogeneous microstructure. Pellets (b) and (c) were irradiated at 430 W/cm for 10 min and 24 h, respectively. The high linear heat rate caused considerable restructuring in the radial direction of the pellets (Figure 5.6b and c). Figure 5.7 shows the microstructure in a part of the cross section [1,3,4]. The unchanged microstructure area, equiaxed grain area, columnar grain area, and central hole exist from the outer periphery to the pellet center as restructured. This restructuring occurs early in the irradiation in the sodium-cooled fast reactor. Fuel pellet burning accumulates fission products (FPs) and changes chemical stability. The irradiation test was conducted as a parameter of the O/M ratio of MOX pellets (Figure 5.7). The effect of the O/M ratio on restructuring was observed. In the irradiation behavior of a MOX pellet with O/M=2.00, the restructuring was observed in a wider area, and a larger diameter of the central hole was observed than the low O/M pellet. Such extensive restructuring at O/M=2.00 could not be represented in the conventional performance analysis code. A physicochemical model is needed to understand and represent various fuel behaviors depending on composition and irradiation conditions.

The maximum fuel temperature limits the maximum linear heat rate in fuel design. Because the temperature and temperature gradient determine the mass transfer causing these irradiation behaviors, the radial temperature distribution of the fuel must be evaluated. The linear heat rate (P) can be described using Eq. (5.23) as a function of λ using the cylindrical coordinate system.

$$P = Q/4 \cdot r_1^2 = \lambda(T_0 - T_1) \tag{5.23}$$

Here, Q is the power per unit volume, r_i is the radial direction distance and T_i is the temperature at r_i.

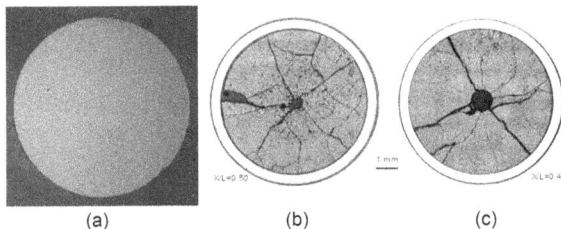

(a)   (b)   (c)

FIGURE 5.6   Microstructure images of the cross section of (a) sintered pellets and (b) irradiated pellets at 430 W/cm for 10 min and (c) 24 h.

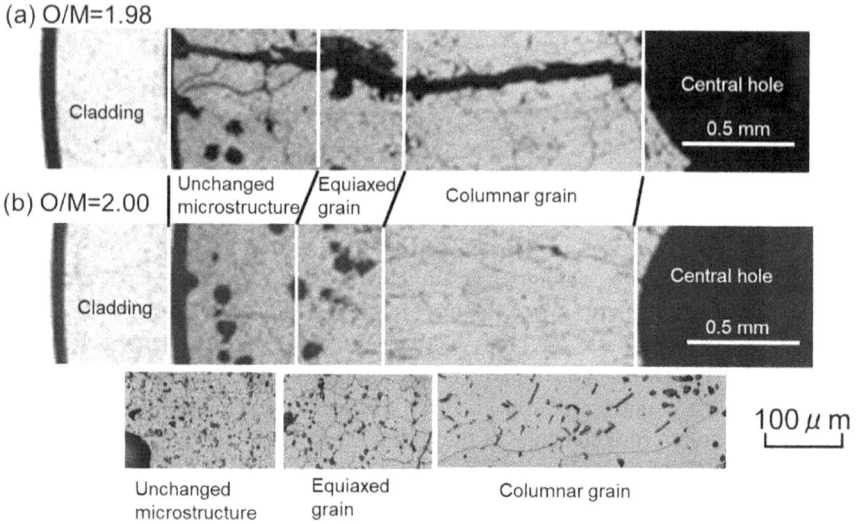

FIGURE 5.7 Part of the microstructure on the cross section of pellets with (a) oxygen-to-metal (O/M) = 1.98 and (b) O/M = 2.00 irradiated at 470 W/cm.

The fuel pin's radial temperature distribution can be calculated from the thermal conductivity, linear heat rate, and dimensions of pellets and cladding related to the coolant temperature. The cladding tube's inner surface temperature T_{c1} can be obtained from the coolant temperature using the following formula if the cladding outer surface temperature T_{c2} equals the coolant temperature.

$$Q = -2 \cdot \pi \cdot \lambda_c (T_{c1} - T_{c2}) \ln\left(\frac{d_{c1}}{d_{c2}}\right) \tag{5.24}$$

λ_c is the cladding's thermal conductivity and d_{c1} and d_{c2} are the cladding tube's inner and outer diameters, respectively.

The outer surface of the fuel pellet T_1 can be obtained from the following equation by using the thermal conductance between the cladding and the pellet h_G, called the gap conductance.

$$Q = h_G \cdot 2 \cdot \pi \cdot r_1 \left(T_p - T_{c1}\right) \tag{5.25}$$

According to Eq. (5.23), the pellet's temperature distribution can be evaluated based on its outer surface temperature T_p.

The cladding tube and pellets' thermal conductivity can be known from the experimental results, but it is challenging to find the gap conductance, and the evaluation is performed analytically. A fuel irradiation test using a thermometer and a part-to-melt irradiation test were conducted to obtain the maximum temperature of the fuel being irradiated. The maximum fuel temperature can be obtained from the thermocouple and the fuel's melting point from these tests. The temperature distribution in the fuel pellet can be evaluated from Eq. (5.23) based on the maximum temperature. Because the cladding tube's inner surface temperature can be obtained using Eq. (5.24), the

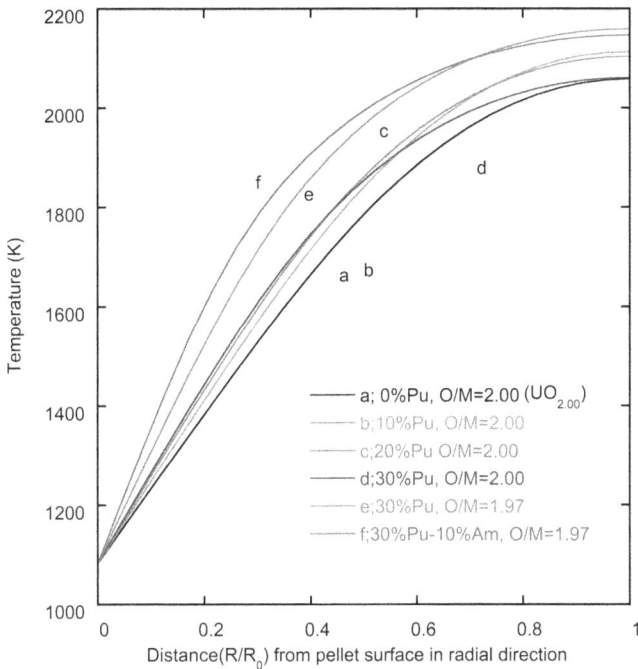

FIGURE 5.8 Temperature profile in the radial direction of pellets with a 95%TD as a composition function.

gap conductance can be estimated from the cladding tube's inner surface temperature and the pellet's surface temperature using Eq. (5.25). Gap conductance can be obtained from the linear heat rate and gap length, as shown in the following equation, by applying Ross & Stoute-type gap conductance [5]. The analytical evaluation of the coefficient gives the following [6]:

$$h_G = C_1 + C_2 \cdot P / \{G_0 - C_3 \cdot d_{c1} \cdot P + C_4\}, \tag{5.26}$$

where h_G is the total heat transfer through the gap (W/cm^2.K), P is the linear heat rating (W/cm), D_g is the average gap diameter (cm), and C_1, C_2, C_3, and C_4 are constant parameters.

A temperature profile was calculated for $P = 380$ W/cm to compare the effect of the fuel's thermal conductivity on temperature distribution, assuming a pellet outer surface temperature of 1,100 K. Figure 5.8 shows the temperature profile's calculation results. The center maximum temperature was 2058.5 K for the UO$_2$ pellet. The temperature increased with the Pu content in the MOX pellets due to the decreased λ. The increased Pu content increased the electronic conduction term λ_e; therefore, the maximum temperature of 30%Pu content MOX was similar to UO$_2$. The O/M decrease and Am content increased the temperature, and the maximum temperature attained to ~2,150 K due to the decreasing λ. In this way, the temperature profile analysis can be performed depending on the composition using the Sci-M Pro. Here, the temperature analysis was performed using only the λ, but by using all the physical properties, it will be possible to evaluate various behaviors, such as Pellet Cladding Mechanical Interaction (PCMI), microstructure change, and redistribution occurring in steady state and transient irradiation. Moreover, knowing the burn up effect on properties is essential to evaluate high burn up fuel performance. Fuel temperature reliability can be improved by iterating the calculated results of density changes and O/M redistribution occurring during irradiation.

5.4   O/M REDISTRIBUTION

Masato Kato, Takayuki Ozawa, and Yoshihisa Ikusawa

Due to the large thermal gradient, O/M redistribution occurs in the pellet's radial direction during irradiation [7–9]. Previous studies have developed models to represent O/M redistribution. Figure 5.9 shows the calculation results of the O/M ratio in the radial direction of $(U_{0.7}Pu_{0.3})O_{2-x}$ pellets, conducted using Aitken and Sari's models depending on the pellet's average O/M ratio. From the figure, the O/M ratio became almost 2.00 and low at the pellet's periphery and center, respectively. Pellets with O/M = 1.98 had a similar redistribution as Aitken and Sari's model. However, Aitken's model showed that the local O/M ratio decreased to 1.82 in the calculation results of MOX with O/M = 1.96. Such a low O/M ratio cannot exist in the phase diagram of $(U_{0.7}Pu_{0.3})O_{2-x}$. Therefore, Sari's model was applied in this work to evaluate the O/M redistribution using the following equations:

$$\ln\left(\frac{x_1}{x_2}\right) = -\frac{Q^*}{R}\left(\frac{1}{T_2} - \frac{1}{T_1}\right) \tag{5.27}$$

$$Q^* = -9.45\times10^5 + 5.66\times10^5 V_{Pu} - 8.5\times10^4 V_{Pu}^{\ 2} \tag{5.28}$$

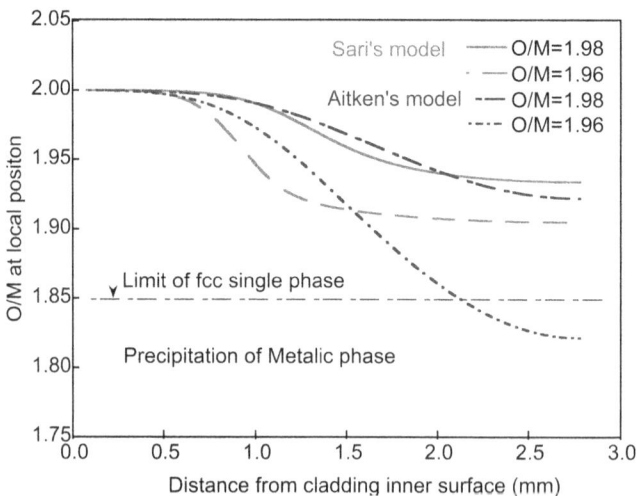

FIGURE 5.9   Calculation results of oxygen-to-metal (O/M) redistribution in the radial direction of pellets.

This calculation result shows the equilibrium condition. The O/M change depending on time is evaluated using the oxygen diffusion coefficient written by

$$J = -ND\left(\nabla c + c\frac{Q}{RT}\nabla T\right). \tag{5.29}$$

Assuming axis symmetry, the following equation can be obtained:

$$\frac{\partial c}{\partial t} = \frac{1}{r}\cdot\frac{\partial}{\partial r}\left\{rD\left(\frac{\partial c}{\partial r} + c\frac{Q}{RT^2}\frac{\partial T}{\partial r}\right)\right\} \tag{5.30}$$

$\dfrac{\partial C}{\partial t} = 0$, when $t \to \infty$. Equation (5.30) is rewritten as

$$\frac{\partial c}{\partial r} = -c\frac{Q}{RT^2}\frac{\partial T}{\partial r}. \tag{5.31}$$

The local O/M ratio can be obtained using

$$c_{\infty, i,j-1} = c_{i,j} + \Delta c_{i,j}, \tag{5.32}$$

$$Q = -8.12\times10^{-4}\exp(4.85\cdot V_{Pu}),\text{ and} \tag{5.33}$$

$$c_{i,j} = c_{\infty} + (c_0(r) - c_{\infty})\cdot\exp(-17.2\widetilde{D_O}(t-t_0)/r_{eq}^2). \tag{5.34}$$

Figure 5.10 shows the calculation results of O/M redistribution in the radial direction of the $(U_{0.7}Pu_{0.3})O_{1.98}$ pellet irradiated at 430 W/cm. Equation (5.6) was used for $\widetilde{D_O}$

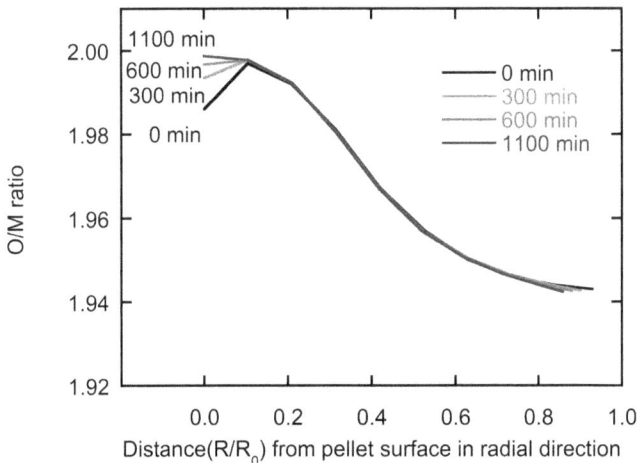

FIGURE 5.10 Oxygen-to-metal (O/M) redistribution of a pellet irradiated at 430 W/cm, calculated using Eq. (5.6).

calculation. The pellet's center and surface temperatures were 2,200 and 1,000 K, respectively. The O/M ratio in the 0–2.5 mm region attained equilibrium during a power increase to 430 W/cm because of the fast oxygen diffusion. We can see oxygen migration around the pellet surface in the figure as a function of time. After 1,000 min, the O/M ratio achieved equilibrium. The O/M ratio redistribution affects the chemical stability of fuels, such as the chemical form of FPs and the chemical interaction between pellet and cladding.

5.5 RESTRUCTURING CAUSED BY VAPOR PRESSURE

Masato Kato, Takayuki Ozawa, Yoshihisa Ikusawa, and Shun Hirooka

The evaporation–agglutination model was applied to represent restructuring in the pellet's radial direction. Vapor pressure is vital in fuel performance to describe mass transfers at high temperatures, and studies are limited because of experimental challenges. In this work, the Rand–Markin model was extended to calculate all vapor species on Am-bearing MOX using Eq. (5.35) [2,8].

$$p(UO_2) = (1 - q - r) \exp\left(-\frac{\Delta G^\circ_{UO_2,Vap.}}{RT}\right)$$

$$p(PuO_2) = q \cdot p_{O_2}^{m/2} \cdot \exp\left(-\frac{\Delta G^\circ_A}{RT}\right)$$

$$\Delta G^\circ_A = -\frac{1}{2}\int_0^m RT \ln p_{O_2} dm' + \Delta G^\circ_{PuO_2,Vap.}$$

$$p(PuO) = p(PuO_2) / \left\{\exp\left(-\frac{\Delta G^\circ_{PuO/PuO_2.}}{RT}\right) \cdot p_{O_2}^{1/2}\right\}$$

$$p(Pu) = p(UO_2) / \left\{\exp\left(-\frac{\Delta G^\circ_{Pu/PuO.}}{RT}\right) \cdot p_{O_2}^{1/2}\right\}$$

$$p(UO) = p(UO_2) / \left\{\exp\left(-\frac{\Delta G^\circ_{UO/UO_2.}}{RT}\right) \cdot p_{O_2}^{1/2}\right\}$$

$$p(U) = P(UO) / \left\{\exp\left(-\frac{\Delta G^\circ_{U/UO.}}{RT}\right) \cdot p_{O_2}^{1/2}\right\}$$

$$p(UO_3) = \exp\left(-\frac{\Delta G^\circ_{UO_2/UO_3.}}{RT}\right) \cdot p_{O_2}^{1/2} \cdot p(UO_2)$$

$$p(AmO_2) = r \cdot p_{O_2}^{n/2} \cdot \exp\left(-\frac{\Delta G^\circ_B}{RT}\right)$$

$$\Delta G_B^\circ = -\frac{1}{2}\int_0^m RT \ln p_{O_2} dn' + \Delta G_{AmO2,Vap.}^\circ$$

$$p(AmO) = p(AmO_2)/\left\{\exp\left(-\frac{\Delta G_{AmO/AmO2.}^\circ}{RT}\right)\cdot p_{O_2}^{1/2}\right\}$$

$$p(Am) = p(UO_2)/\left\{\exp\left(-\frac{\Delta G_{Am/AmO.}^\circ}{RT}\right)\cdot p_{O_2}^{1/2}\right\}$$

$$\Delta G_i^\circ = 257,000 - 68T \text{ for } \frac{1}{2}O_2 = O$$

$$\Delta G_i^\circ = -528,000 + 62T \text{ for } U + \frac{1}{2}O_2 = UO$$

$$\Delta G_i^\circ = -471,000 + 71T \text{ for } UO + \frac{1}{2}O_2 = UO_2$$

$$\Delta G_i^\circ = -44,000 + 90T \text{ for } UO_2 + \frac{1}{2}O_2 = UO_3$$

$$\Delta G_i^\circ = -498,000 + 46T \text{ for } Pu + \frac{1}{2}O_2 = PuO$$

$$\Delta G_i^\circ = -352,000 + 69T \text{ for } PuO + \frac{1}{2}O_2 = PuO_2$$

$$\Delta G_i^\circ = -250,600 + 59.6T \text{ for } AmO + \frac{1}{2}O_2 = AmO_2$$

$$\Delta G_i^\circ = 571,000 + 150T \text{ for } UO_2(s) = UO_2(g)$$

$$\Delta G_i^\circ = 571,000 + 150T \text{ for } PuO_2(s) = PuO_2(g)$$

$$\Delta G_i^\circ = 541,000 + 139.4T \text{ for } AmO_2(s) = AmO_2(g)$$

$$\Delta \bar{G}(PuO_2,s) = -1,047,800 + 187.71T$$

$$\Delta G_f^\circ(PuO_2,g) = -426,050 - 2.96T$$

$$\Delta G_f^\circ(PuO,g) = -74,325 - 72.80T$$

$$\Delta G_f^\circ (Pu, g) = 34,010 - 106.0T$$

$$\Delta \bar{G} (AmO_2, s) = -1,765,000 + 329.01T$$

$$\Delta G_f^\circ (AmO_2, g) = -388,317 - 30,668T$$

$$\Delta G_f^\circ (AmO, g) = -137,707 - 28.772T$$

$$\Delta G_f^\circ (Am, g) = 262,437 - 111.41T$$

$$\Delta \bar{G} (UO_2, s) = -1,080,000 + 169.0T$$

$$\Delta G_f^\circ (UO_2, g) = -502,800 + 9.06T$$

$$\Delta G_f^\circ (UO_3, g) = -809,310 + 68.06T$$

$$\Delta G_f^\circ (UO, g) = -16,550 - 73T$$

$$\Delta G_f^\circ (U, g) = 500,360 - 120.2T \tag{5.35}$$

Equation (5.4) was used to calculate P_{O_2}. Figure 5.11 shows the O/M dependence of vapor pressure at 2,000 K. The vapor pressure of UO_3 rapidly increases around O/M = 2.00 and is dominant in the O/M range above 2.00. In the low O/M ratio regions, the vapor pressure of AmO is the highest among all species.

Pore migration toward the pellet center restructures the radial direction of the pellets (Figure 5.7) [3,4]. The pore migration rate V is represented by the following equation:

$$V = \Omega D \frac{d}{dT} \left(\frac{p}{kT} \right) \left| \frac{dT}{dr} \right|, \tag{5.36}$$

where Ω is the molecular volume, and D is the gas diffusion coefficient.

$$\left(\frac{dT}{dr} \right)_{pore} = \alpha \left(\frac{dT}{dr} \right)_{Matrix} \tag{5.37}$$

It was reported that α is 1.5–4. Figure 5.12 shows the vapor pressure of all species on MOX with O/M = 2.00 and 1.96 in the pellet's radial direction, and Figure 5.13 shows the temperature profile. The temperature of MOX pellets with O/M = 2.00 is lower

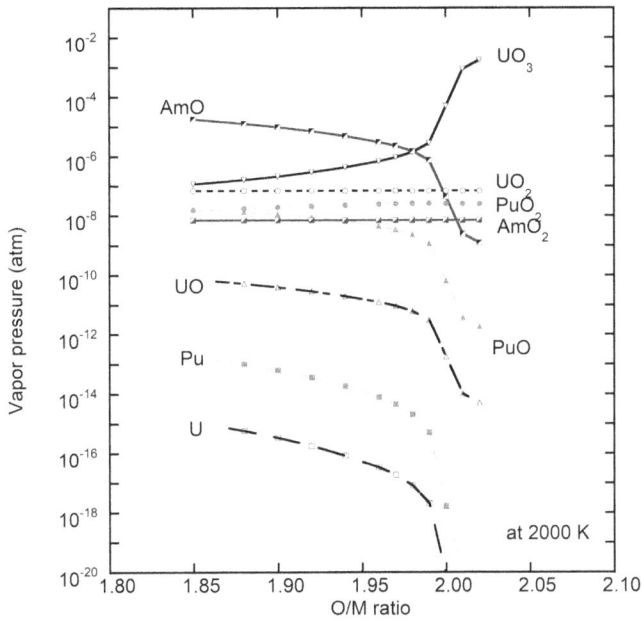

FIGURE 5.11   Oxygen-to-metal (O/M) dependence of vapor pressure on $(U_{0.65}Pu_{0.3}Am_{0.05})$ O_{2-x} at 2,000 K.

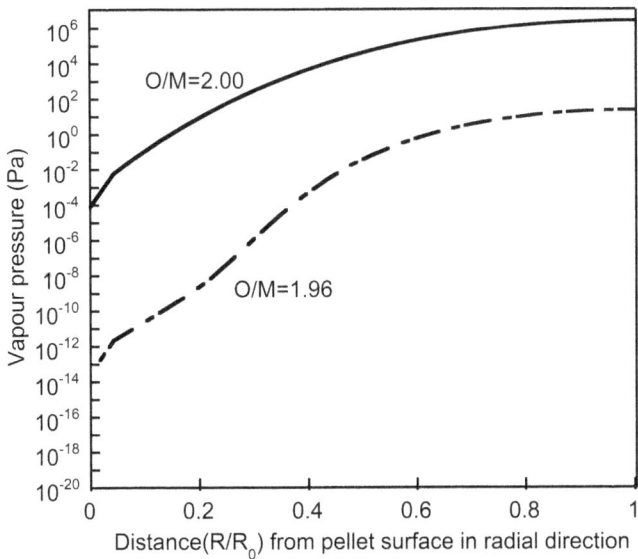

FIGURE 5.12   Vapor pressure in the radial direction of the pellet.

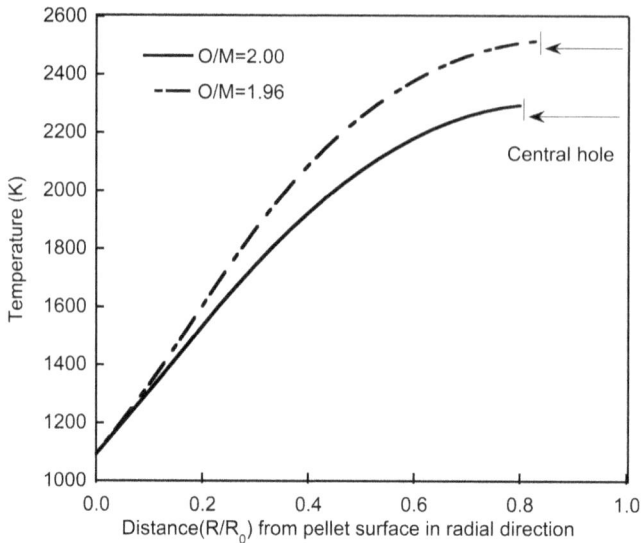

FIGURE 5.13   Temperature profile in the radial direction of the pellets.

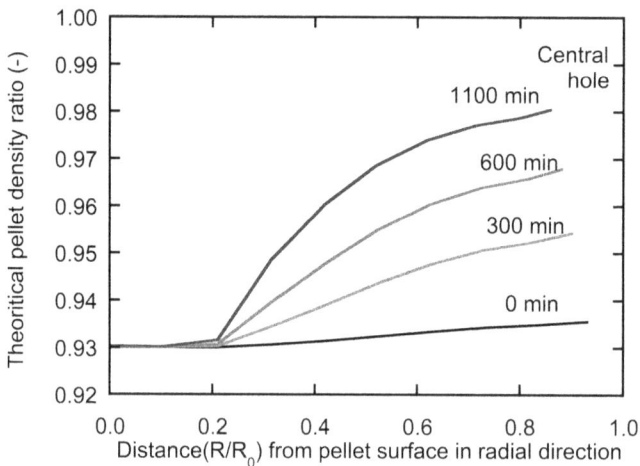

FIGURE 5.14   Change in the theoretical pellet density ratio during irradiation at 430 W/cm.

than the low O/M pellets because of high thermal conductivity. The vapor pressure on MOX pellets with O/M = 2.00 was higher than low O/M pellets due to the vapor pressure of UO_3. The high vapor pressure increased the central void diameter. By calculating the vapor pressure of all species, this model could represent the density change in the radial direction depending on the O/M ratio. Figure 5.14 shows a pellet density change depending on when the linear heating rate was 430 W/cm. The pellet density and the central hole diameter increase with irradiated time.

5.6   CHEMICAL STABILITY OF FISSION PRODUCTS

Several FPs are generated and accumulated during burning. The fission of a Pu or U atom generates two FPs and oxygen liberated from MO_2. These elements are present in the fuel, depending on their chemical properties. FPs can be classified into four types according to their chemical behavior [7,8].

Type 1: gases and volatile elements (Xe, Kr, Br, I)
Type 2: FPs forming metallic precipitates (Tc, Pd, Rd, Ru, Ir, Ag)
Type 3: FPs forming oxide precipitates (Rb, Cs, Ba, Zr)
Type 4: FPs dissolved in U and Pu sites (Nd, La, Ce, Zr)

The behavior of type 1 elements, Xe and Kr, is vital in fuel design. They release from the pellet to the free volume inside the fuel pin and increase the inner pressure and PCI. The fuel pellet and FPs' chemical stability affect the fuel behavior. MOX pellets with O/M = 1.98–1.97 are produced and loaded as fresh fuel. An average O/M ratio of MOX pellet approaches O/M = 2.00 due to the chemical stability changes during burning. Figure 5.15 shows the oxygen potentials of MOX as functions of

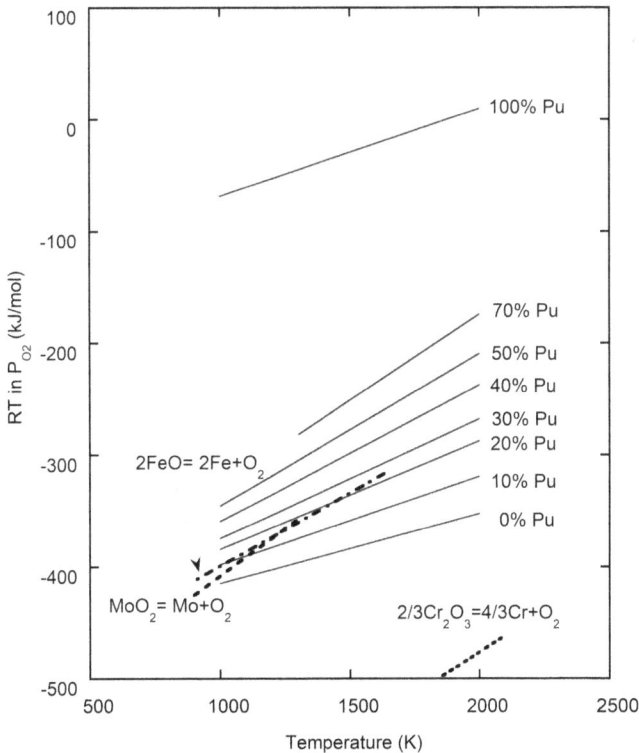

FIGURE 5.15 $\Delta \overline{G_{O_2}}$ change in $(U, Pu)O_{2.00}$ depending on the Pu content in an Ellingham diagram.

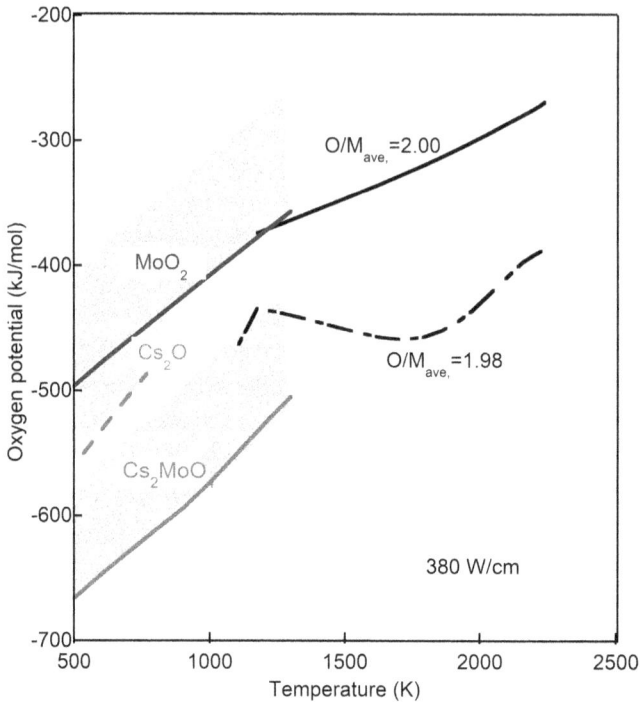

FIGURE 5.16 $\Delta \overline{G}_{O_2}$ at a local position in the radial direction of the $(U_{0.8}Pu_{0.2})O_{2.00}$ pellet with an average oxygen-to-metal $(O/M) = 2.00$ and 1.98.

temperature and Pu content. $\Delta \overline{G}_{O_2}$ of the MOX pellet increases with Pu content. Section 5.4 describes the O/M redistribution in the MOX pellet's radial direction. The redistribution was calculated in $(Pu_{0.8}Pu_{0.2})O_{1.98}$ irradiated at 380 W/cm, and $\Delta \overline{G}_{O_2}$ was evaluated at the local position in the pellet's radial direction. Figure 5.16 shows the evaluated data.

Mo is a central FP, precipitated as metal or oxide inside MOX pellets. Figures 5.15 and 5.16 show the G° of $MoO_2 \leftrightarrow Mo + O_2$. The G° is approximately 400 kJ/mol. The figures show that MoO_2 precipitates from the pellet's outer periphery at $O/M = 2.00$, and the increase in $\Delta \overline{G}_{O_2}$ of MOX is suppressed because of MoO_2 formation. Cs_2MoO_4 is familiar as precipitates at the pellet's periphery during burning, the Joint-Oxyde-Gaine (JOG) phase [10]. Figure 5.16 shows the G° of the JOG phase, showing that the JOG phase also precipitates from the pellet's outer periphery with an average O/M ratio of 1.98. The JOG phase has lower thermal conductivity than MOX, so the fuel temperature increases.

G° of $Fe + 1/2\ O_2 \leftrightarrow FeO$ would cause homogeneous oxidation on the cladding's inner surface. $a = 4/\sqrt{3} \cdot \{r_c (1 + 0.112x) + r_a\}$ of MOX pellets increasing with burning prompts cladding oxidation, increasing the cladding's creep stress because of the reduction in cladding wall thickness. This phenomenon will limit fuel life. If the chemical stability in the fuel can be evaluated in this way, it will be possible to evaluate the chemical form of FP and the oxidation behavior of the cladding tube.

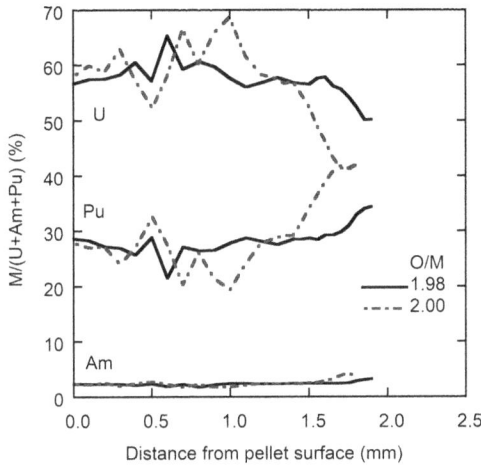

FIGURE 5.17   Redistribution of actinide elements in a radial direction of the pellets.

5.7   OTHER FUEL BEHAVIORS

Masato Kato, Shinji Sasaki, and Koji Maeda

Understanding and describing various irradiation behaviors is crucial for evaluating fuel performance. The main irradiation behaviors, excluding those described in precious sections are outlined. When the operation of fresh fuels is started, tensile stress in the fuel pellet's circumferential direction increases due to the pellet's thermal expansion. The pellet breaks into multiple pieces. The increase in fuel temperature shrinks the pellet's sintering and widens its gap. The pellet pieces are then relocated to fill the gap. Furthermore, the cracks generated at the initial irradiation stage heal and disappear in the high-temperature region observed in the early irradiation stage and affect the fuel temperature. Simultaneously, the restructuring in the pellet's radial direction is also formed, such as grain growth, columnar grain, and central hole, as shown in the former sections. The redistribution of actinides and oxygen also occurs. Figure 5.17 shows the element analysis results in the radial direction of the fuel pellets with (a) O/M = 2.00 and (b) O/M = 1.98 [1,2,11,12]. The Pu and Am concentrations increase around the central hole. In the fuel design of fast reactors, the fuel's melting point limits the maximum linear heat rate. Because such a change in the actinide concentration lowers the melting point and the limit value of the linear heat power, its evaluation is vital. The figure shows that a significant redistribution occurs in pellets with O/M = 2.00 compared with low O/M MOX pellets, and it is necessary to consider the effect of vapor pressure to evaluate this phenomenon.

When the burn up increases, the accumulation of FPs changes the fuel's chemical stability, and the FP gas moves to the grain boundaries to form pores. It is released to the pellets outside. The fuel design is conducted for fast reactor fuel, assuming a 100% FP gas release. The pore formation caused by FP gas will reduce fuel pellet's density and affect PCMI. The creep strength of the cladding tube limits the life of the burning fuel.

Irradiation reduces creep strength, and chemical interaction with pellets increases stress due to the thinning of the cladding wall due to cladding oxidation, limiting fuel life. The behaviors shown here are only parts, but it is crucial to express various fuel behaviors using a physicochemical model for fuel performance and safety evaluation.

REFERENCES

1. Maeda, K., et al., Short-term irradiation behavior of minor actinide doped uranium plutonium mixed oxide fuels irradiated in an experimental fast reactor. *Journal of Nuclear Materials*, 2009. **385**(2): pp. 413–418.
2. Maeda, K., et al., Radial redistribution of actinides in irradiated FR-MOX fuels. *Journal of Nuclear Materials*, 2009. **389**(1): pp. 78–84.
3. Ozawa, T., et al., Development of fuel performance analysis code, BISON for MOX, named Okami: Analyses of pore migration behavior to affect the MA-bearing MOX fuel restructuring. *Journal of Nuclear Materials*, 2021. **553**: p. 153038.
4. Ikusawa, Y., et al., Oxide-metal ratio dependence of central void formation of mixed oxide fuel irradiated in fast reactors. *Nuclear Technology*, 2017. **199**(1): pp. 83–95.
5. Ross, A.M. and R.L. Stoute, *Heat Transfer Coefficient between UO₂ and Zircaloy-2*. 1962, Atomic Energy of Canada, Chalk River, ON.
6. Ikusawa, Y., S. Hirooka, and M. Uno. Oxygen potential and self-irradiation effects on fuel temperature in Am-MOX, in *Generation IV International Forum*. 2018, Nuclear Energy Agency, Paris.
7. Kleykamp, H., The chemical state of the fission products in oxide fuels. *Journal of Nuclear Materials*, 1985. **131**(2–3): pp. 221–246.
8. Olander, D.R., *Fundamental Aspects of Nuclear Reactor Fuel Elements*. 1976, Technical Information Center, Office of Public Affairs, Energy Research and Development Administration, Oak Ridge, TN.
9. Sari, C. and G. Schumacher, Oxygen redistribution in fast reactor oxide fuel. *Journal of Nuclear Materials*, 1976. **61**(2): pp. 192–202.
10. Cappia, F., et al., Post-irradiation examinations of annular mixed oxide fuels with average burnup 4 and 5% FIMA. *Journal of Nuclear Materials*, 2020. **533**: p. 152076.
11. Tanaka, K., et al., Restructuring and redistribution of actinides in Am-MOX fuel during the first 24h of irradiation. *Journal of Nuclear Materials*, 2013. **440**(1–3): pp. 480–488.
12. Sasaki, S. and K. Maeda. Irradiation behaviour of Am bearing MOX fuel, in *15th International Exchange Meeting on Actinide and Fission Product Partitioning and Transmutation*. 2018, OECD/NEA, Manchester, UK.

6 Science-Based Physical Property Model and Outlook for Fuel Development

In this work, a database on mixed oxide (MOX) properties, including oxygen potential, melting point, lattice parameter, thermal expansion, sound speeds, thermal diffusivity, oxygen self-diffusion diffusion coefficient, and chemical diffusion coefficient, was set using experimental data. The relationship among properties included in the database was evaluated, and a science-based model of MOX properties was derived for advanced fuel technologies. This model can describe property changes as functions of Pu content, minor actinide (MA) content, oxygen-to-metal (O/M) ratio, and temperatures. Oxygen potential data were evaluated based on the defect chemistry. The relational correlation representing the oxygen potential was expanded to determine the relationship of $P_{O_2} - O/M - T$ in all regions of 0%–100% Pu content. Furthermore, the effect of MA addition was included.

The relational correlation was analyzed to construct Brouwer's diagram and calculate the point defect concentrations of $\left[Vo^{..} \right]$, $\left[Oi'' \right]$, $\left[e' \right]$, and $\left[h^{.} \right]$. The calculation results showed that $\left[e' \right]$ and $\left[h^{.} \right]$ have maximum values at approximately 50% Pu in the UO_2–PuO_2 system. The increase in electronic concentration increased the electrical conductivity, linked to electronic defect and electrical-conduction contributions at high temperatures in heat capacity and thermal conductivity, respectively. Furthermore, the evaluation results of $\left[Vo^{..} \right]$ and $\left[Oi'' \right]$, depending on P_{O_2}, were used to describe oxygen self-diffusion and chemical diffusion coefficients. The analysis result showed that the diffusion coefficients changed near the stoichiometric region, and the diffusion of Oi' was dominated around the stoichiometric composition.

Oxygen potential was extrapolated up to the melting temperature, obtaining the relationship between oxygen potential and solidus temperature. Solidus temperature can be expressed as Pu, MA, and O/M dependence functions. It was estimated that MA content decreases solidus temperature by 6 K–7 K and 1 K–2 K per 1% Am and 1% Np content, respectively.

The heat capacity was represented as $C_p = C_l + C_d + C_{sch} + C_{exe}$. Only Schottky-type heat capacity was obtained using the ab-initio approach. Thermal conductivity comprised two contributions: phonon- and electrical-conduction mechanisms. Other parameters for calculating heat capacity and thermal conductivity were obtained from relational correlations for lattice parameters, thermal conductivity, and sound speeds. The analysis provided a crucial perspective on mechanisms exited at high

DOI: 10.1201/9781003298205-6   **127** heat capacity and thermal conductivity temperatures. Using Kopp's law, the MOX heat capacity was evaluated from UO_2 and PuO_2. However, at temperatures higher than 1,500 K, the results showed that Kopp's law could not represent the MOX heat capacity, and it was higher at high temperatures than that of UO_2 and PuO_2. The evaluation results of thermal conductivity showed that the electrical-conduction term of MOX increased.

In this study, a science-based model to represent MOX properties was derived to extrapolate the properties in a wide composition range. This model was applied to the fuel-performance code and simulated irradiation behavior. It can analyze various oxide fuels, such as fast reactor, light water reactor, and MA content fuel. The current model could be used in a performance code to analyze various oxide fuels' fuel safety and performance. However, there are few experimental data in the high temperature region above 2,000K where the Bredig transition occurs, and the calculation results of the physical property model are not sufficient. Computational science work was described in Chapter 4, but we have been developing a machine learning molecular dynamics simulation to evaluate the basic physical properties of advanced nuclear fuel. This simulation can cover properties in areas where experimental data is lacking and can expand the database in addition to experimental data. In the future, we believe that technology for evaluating irradiation performance using machine learning models that apply data science will be developed.

Appendixes Database for MOX Properties

TABLE 1

Oxygen Potential

U	Pu	Am	Np	O/M	Temp. (K)	$\Delta \bar{G}_{O_2}$ (kJ/mol)
0.623	0.350	0.027	0.000	2.0015	1,673	−223.5
0.623	0.350	0.027	0.000	2.0003	1,673	−259.3
0.623	0.350	0.027	0.000	2.0001	1,673	−277.6
0.623	0.350	0.027	0.000	2.0000	1,673	−252.0
0.623	0.350	0.027	0.000	1.9999	1,673	−288.9
0.623	0.350	0.027	0.000	1.9997	1,673	−305.5
0.623	0.350	0.027	0.000	1.9990	1,673	−323.1
0.623	0.350	0.027	0.000	1.9988	1,673	−339.7
0.623	0.350	0.027	0.000	1.9987	1,673	−314.1
0.623	0.350	0.027	0.000	1.9987	1,673	−341.3
0.623	0.350	0.027	0.000	1.9984	1,673	−360.9
0.623	0.350	0.027	0.000	1.9978	1,673	−362.3
0.623	0.350	0.027	0.000	1.9970	1,673	−378.9
0.623	0.350	0.027	0.000	1.9968	1,673	−383.5
0.623	0.350	0.027	0.000	1.9960	1,673	−386.2
0.623	0.350	0.027	0.000	1.9936	1,673	−398.1
0.623	0.350	0.027	0.000	1.9933	1,673	−401.1
0.623	0.350	0.027	0.000	1.9910	1,673	−408.1
0.623	0.350	0.027	0.000	1.9885	1,673	−415.0
0.623	0.350	0.027	0.000	1.9855	1,673	−427.0
0.623	0.350	0.027	0.000	2.0031	1,773	−206.0
0.623	0.350	0.027	0.000	2.0013	1,773	−218.3
0.623	0.350	0.027	0.000	2.0008	1,773	−238.0
0.623	0.350	0.027	0.000	2.0005	1,773	−251.7
0.623	0.350	0.027	0.000	1.9999	1,773	−270.7
0.623	0.350	0.027	0.000	1.9997	1,773	−279.2
0.623	0.350	0.027	0.000	1.9995	1,773	−301.3
0.623	0.350	0.027	0.000	1.9989	1,773	−321.4
0.623	0.350	0.027	0.000	1.9985	1,773	−329.1
0.623	0.350	0.027	0.000	1.9984	1,773	−327.0
0.623	0.350	0.027	0.000	1.9980	1,773	−327.7

(Continued)

TABLE 1 (*Continued*)
Oxygen Potential

U	Pu	Am	Np	O/M	Temp. (K)	$\Delta \bar{G}_{O_2}$ (kJ/mol)
0.623	0.350	0.027	0.000	1.9978	1,773	−339.0
0.623	0.350	0.027	0.000	1.9964	1,773	−343.2
0.623	0.350	0.027	0.000	1.9950	1,773	−370.3
0.623	0.350	0.027	0.000	1.9950	1,773	−360.4
0.623	0.350	0.027	0.000	1.9940	1,773	−362.2
0.623	0.350	0.027	0.000	1.9901	1,773	−369.2
0.623	0.350	0.027	0.000	1.9885	1,773	−382.9
0.623	0.350	0.027	0.000	1.9880	1,773	−382.6
0.623	0.350	0.027	0.000	1.9855	1,773	−394.9
0.623	0.350	0.027	0.000	1.9774	1,773	−416.7
0.623	0.350	0.027	0.000	1.9771	1,773	−396.3
0.623	0.350	0.027	0.000	1.9722	1,773	−418.5
0.623	0.350	0.027	0.000	2.0040	1,873	−191.3
0.623	0.350	0.027	0.000	2.0039	1,873	−189.0
0.623	0.350	0.027	0.000	2.0010	1,873	−220.6
0.623	0.350	0.027	0.000	2.0008	1,873	−230.3
0.623	0.350	0.027	0.000	2.0002	1,873	−258.9
0.623	0.350	0.027	0.000	2.0002	1,873	−263.0
0.623	0.350	0.027	0.000	1.9995	1,873	−277.9
0.623	0.350	0.027	0.000	1.9994	1,873	−280.8
0.623	0.350	0.027	0.000	1.9993	1,873	−280.5
0.623	0.350	0.027	0.000	1.9990	1,873	−302.0
0.623	0.350	0.027	0.000	1.9988	1,873	−301.3
0.623	0.350	0.027	0.000	1.9977	1,873	−300.9
0.623	0.350	0.027	0.000	1.9977	1,873	−280.5
0.623	0.350	0.027	0.000	1.9972	1,873	−317.3
0.623	0.350	0.027	0.000	1.9971	1,873	−316.5
0.623	0.350	0.027	0.000	1.9966	1,873	−313.9
0.623	0.350	0.027	0.000	1.9965	1,873	−321.7
0.623	0.350	0.027	0.000	1.9961	1,873	−321.4
0.623	0.350	0.027	0.000	1.9958	1,873	−328.1
0.623	0.350	0.027	0.000	1.9937	1,873	−336.6
0.623	0.350	0.027	0.000	1.9929	1,873	−341.4
0.623	0.350	0.027	0.000	1.9920	1,873	−345.1
0.623	0.350	0.027	0.000	1.9860	1,873	−362.2
0.623	0.350	0.027	0.000	1.9848	1,873	−372.3
0.623	0.350	0.027	0.000	1.9847	1,873	−367.4
0.623	0.350	0.027	0.000	1.9827	1,873	−371.5
0.623	0.350	0.027	0.000	1.9728	1,873	−390.1

(Continued)

TABLE 1 (Continued)
Oxygen Potential

U	Pu	Am	Np	O/M	Temp. (K)	$\Delta \bar{G}_{O_2}$ (kJ/mol)
0.623	0.350	0.027	0.000	1.9653	1,873	−403.1
0.623	0.350	0.027	0.000	1.9603	1,873	−403.5
0.553	0.285	0.015	0.147	2.0046	1,873	−227.2
0.553	0.285	0.015	0.147	2.0043	1,873	−196.4
0.553	0.285	0.015	0.147	2.0015	1,873	−257.1
0.553	0.285	0.015	0.147	2.0014	1,873	−253.8
0.553	0.285	0.015	0.147	2.0009	1,873	−255.8
0.553	0.285	0.015	0.147	2.0005	1,873	−275.9
0.553	0.285	0.015	0.147	2.0001	1,873	−286.8
0.553	0.285	0.015	0.147	1.9988	1,873	−297.0
0.553	0.285	0.015	0.147	1.9985	1,873	−293.2
0.553	0.285	0.015	0.147	1.9970	1,873	−310.1
0.553	0.285	0.015	0.147	1.9954	1,873	−332.7
0.553	0.285	0.015	0.147	1.9886	1,873	−349.8
0.553	0.285	0.015	0.147	1.9880	1,873	−352.0
0.553	0.285	0.015	0.147	1.9788	1,873	−377.1
0.553	0.285	0.015	0.147	1.9751	1,873	−395.0
0.553	0.285	0.015	0.147	1.9620	1,873	−408.9
0.553	0.285	0.015	0.147	1.9597	1,873	−422.5
0.553	0.285	0.015	0.147	1.9456	1,873	−431.8
0.553	0.285	0.015	0.147	1.9362	1,873	−450.4
0.553	0.285	0.015	0.147	2.0057	1,923	−192.2
0.553	0.285	0.015	0.147	2.0025	1,923	−214.5
0.553	0.285	0.015	0.147	2.0006	1,923	−248.9
0.553	0.285	0.015	0.147	2.0000	1,923	−256.2
0.553	0.285	0.015	0.147	2.0000	1,923	−270.2
0.553	0.285	0.015	0.147	1.9991	1,923	−279.5
0.553	0.285	0.015	0.147	1.9990	1,923	−282.2
0.553	0.285	0.015	0.147	1.9972	1,923	−305.9
0.553	0.285	0.015	0.147	1.9967	1,923	−307.9
0.553	0.285	0.015	0.147	1.9944	1,923	−321.2
0.553	0.285	0.015	0.147	1.9903	1,923	−342.3
0.553	0.285	0.015	0.147	1.9889	1,923	−346.3
0.553	0.285	0.015	0.147	1.9886	1,923	−348.0
0.553	0.285	0.015	0.147	1.9850	1,923	−354.3
0.553	0.285	0.015	0.147	1.9784	1,923	−373.7
0.553	0.285	0.015	0.147	1.9762	1,923	−375.6
0.553	0.285	0.015	0.147	1.9704	1,923	−388.9
0.553	0.285	0.015	0.147	1.9670	1,923	−395.3
0.553	0.285	0.015	0.147	1.9500	1,923	−407.3

(Continued)

TABLE 1 (*Continued*)
Oxygen Potential

U	Pu	Am	Np	O/M	Temp. (K)	$\Delta \bar{G}_{O_2}$ (kJ/mol)
0.553	0.285	0.015	0.147	1.9481	1,923	−422.6
0.553	0.285	0.015	0.147	1.9315	1,923	−430.9
0.553	0.285	0.015	0.147	1.9229	1,923	−450.1
0.688	0.290	0.012	0.010	2.0000	1,623	−283.1
0.688	0.290	0.012	0.010	1.9997	1,623	−347.5
0.688	0.290	0.012	0.010	1.9993	1,623	−364.3
0.688	0.290	0.012	0.010	2.0000	1,623	−279.9
0.688	0.290	0.012	0.010	1.9986	1,623	−373.3
0.688	0.290	0.012	0.010	1.9983	1,623	−381.0
0.688	0.290	0.012	0.010	2.0000	1,623	−280.6
0.688	0.290	0.012	0.010	1.9979	1,623	−386.2
0.688	0.290	0.012	0.010	2.0000	1,623	−312.1
0.688	0.290	0.012	0.010	1.9986	1,623	−381.6
0.688	0.290	0.012	0.010	1.9976	1,623	−390.0
0.688	0.290	0.012	0.010	2.0000	1,623	−309.5
0.688	0.290	0.012	0.010	1.9972	1,623	−395.2
0.688	0.290	0.012	0.010	1.9968	1,623	−398.4
0.688	0.290	0.012	0.010	2.0000	1,623	−312.1
0.688	0.290	0.012	0.010	1.9964	1,623	−399.7
0.688	0.290	0.012	0.010	2.0000	1,623	−324.3
0.688	0.290	0.012	0.010	1.9965	1,623	−404.8
0.688	0.290	0.012	0.010	1.9923	1,623	−422.2
0.688	0.290	0.012	0.010	2.0000	1,623	−319.8
0.688	0.290	0.012	0.010	1.9968	1,623	−400.3
0.688	0.290	0.012	0.010	1.9934	1,623	−419.0
0.688	0.290	0.012	0.010	2.0000	1,623	−326.9
0.688	0.290	0.012	0.010	1.9980	1,623	−397.1
0.688	0.290	0.012	0.010	1.9959	1,623	−412.5
0.688	0.290	0.012	0.010	2.0000	1,573	−297.7
0.688	0.290	0.012	0.010	1.9997	1,573	−353.2
0.688	0.290	0.012	0.010	1.9995	1,573	−368.8
0.688	0.290	0.012	0.010	2.0000	1,573	−295.2
0.688	0.290	0.012	0.010	1.9991	1,573	−377.6
0.688	0.290	0.012	0.010	1.9990	1,573	−384.4
0.688	0.290	0.012	0.010	2.0000	1,573	−298.3
0.688	0.290	0.012	0.010	1.9987	1,573	−390.1
0.688	0.290	0.012	0.010	1.9987	1,573	−394.4
0.688	0.290	0.012	0.010	2.0000	1,573	−300.2
0.688	0.290	0.012	0.010	1.9982	1,573	−398.2
0.688	0.290	0.012	0.010	1.9982	1,573	−401.3

(Continued)

TABLE 1 (*Continued*)
Oxygen Potential

U	Pu	Am	Np	O/M	Temp. (K)	$\Delta \bar{G}_{O_2}$ (kJ/mol)
0.688	0.290	0.012	0.010	2.0000	1,573	−299.6
0.688	0.290	0.012	0.010	1.9983	1,573	−398.2
0.688	0.290	0.012	0.010	1.9981	1,573	−401.9
0.688	0.290	0.012	0.010	2.0000	1,573	−310.2
0.688	0.290	0.012	0.010	1.9982	1,573	−401.9
0.688	0.290	0.012	0.010	1.9981	1,573	−404.4
0.688	0.290	0.012	0.010	2.0000	1,573	−313.9
0.688	0.290	0.012	0.010	1.9976	1,573	−404.4
0.688	0.290	0.012	0.010	1.9973	1,573	−406.3
0.688	0.290	0.012	0.010	2.0000	1,573	−317.1
0.688	0.290	0.012	0.010	2.0000	1,573	−300.2
0.688	0.290	0.012	0.010	1.9984	1,573	−395.1
0.688	0.290	0.012	0.010	1.9978	1,573	−406.3
0.688	0.290	0.012	0.010	1.9944	1,573	−426.9
0.688	0.290	0.012	0.010	2.0000	1,573	−319.6
0.688	0.290	0.012	0.010	1.9931	1,573	−427.5
0.688	0.290	0.012	0.010	2.0000	1,573	−323.9
0.688	0.290	0.012	0.010	1.9978	1,573	−406.9
0.688	0.290	0.012	0.010	1.9943	1,573	−426.3
0.688	0.290	0.012	0.010	2.0000	1,573	−320.2
0.688	0.290	0.012	0.010	1.9988	1,573	−393.8
0.688	0.290	0.012	0.010	1.9947	1,573	−423.8
0.688	0.290	0.012	0.010	1.9949	1,573	−420.6
0.688	0.290	0.012	0.010	1.9976	1,573	−405.7
0.688	0.290	0.012	0.010	2.0000	1,573	−325.8
0.688	0.290	0.012	0.010	2.0000	1,473	−313.0
0.688	0.290	0.012	0.010	2.0000	1,473	−363.3
0.688	0.290	0.012	0.010	2.0000	1,473	−322.4
0.688	0.290	0.012	0.010	1.9998	1,473	−363.9
0.688	0.290	0.012	0.010	1.9998	1,473	−377.3
0.688	0.290	0.012	0.010	2.0000	1,473	−318.3
0.688	0.290	0.012	0.010	1.9996	1,473	−384.3
0.688	0.290	0.012	0.010	1.9995	1,473	−391.9
0.688	0.290	0.012	0.010	2.0000	1,473	−311.9
0.688	0.290	0.012	0.010	1.9995	1,473	−391.3
0.688	0.290	0.012	0.010	2.0000	1,473	−311.3
0.688	0.290	0.012	0.010	1.9995	1,473	−400.1
0.688	0.290	0.012	0.010	2.0000	1,473	−312.5
0.688	0.290	0.012	0.010	1.9995	1,473	−400.1
0.688	0.290	0.012	0.010	1.9995	1,473	−403.6

(*Continued*)

TABLE 1 (*Continued*)
Oxygen Potential

U	Pu	Am	Np	O/M	Temp. (K)	$\Delta \bar{G}_{O_2}$ (kJ/mol)
0.688	0.290	0.012	0.010	2.0000	1,473	−314.2
0.688	0.290	0.012	0.010	1.9991	1,473	−407.1
0.688	0.290	0.012	0.010	2.0000	1,473	−319.5
0.688	0.290	0.012	0.010	1.9991	1,473	−411.8
0.688	0.290	0.012	0.010	1.9984	1,473	−431.1
0.688	0.290	0.012	0.010	2.0000	1,473	−338.2
0.688	0.290	0.012	0.010	1.9992	1,473	−406.5
0.688	0.290	0.012	0.010	2.0000	1,273	−326.3
0.688	0.290	0.012	0.010	1.9995	1,273	−407.1
0.688	0.290	0.012	0.010	1.9994	1,273	−417.2
0.688	0.290	0.012	0.010	2.0000	1,273	−331.9
0.688	0.290	0.012	0.010	1.9994	1,273	−407.6
0.688	0.290	0.012	0.010	2.0000	1,273	−328.4
0.688	0.290	0.012	0.010	1.9999	1,273	−380.4
0.688	0.290	0.012	0.010	1.9997	1,273	−393.0
0.688	0.290	0.012	0.010	2.0000	1,273	−326.9
0.688	0.290	0.012	0.010	2.0000	1,273	−337.5
0.688	0.290	0.012	0.010	1.9995	1,273	−413.7
0.688	0.290	0.012	0.010	1.9994	1,273	−423.8
0.688	0.290	0.012	0.010	2.0000	1,273	−342.0
0.688	0.290	0.012	0.010	1.9994	1,273	−421.8
0.688	0.290	0.012	0.010	2.0000	1,273	−345.0
0.880	0.120	–	–	1.9963	1,673	−446.0
0.880	0.120	–	–	1.9967	1,673	−441.4
0.880	0.120	–	–	1.9968	1,673	−440.7
0.880	0.120	–	–	1.9976	1,673	−424.1
0.880	0.120	–	–	1.9975	1,673	−419.5
0.880	0.120	–	–	1.9976	1,673	−412.8
0.880	0.120	–	–	1.9991	1,673	−388.9
0.880	0.120	–	–	2.0001	1,673	−344.4
0.880	0.120	–	–	2.0006	1,673	−321.9
0.880	0.120	–	–	2.0004	1,673	−311.2
0.880	0.120	–	–	2.0009	1,673	−308.6
0.880	0.120	–	–	2.0009	1,673	−292.7
0.880	0.120	–	–	2.0014	1,673	−281.4
0.880	0.120	–	–	2.0013	1,673	−276.7
0.880	0.120	–	–	2.0015	1,673	−272.7
0.880	0.120	–	–	2.0022	1,673	−261.4
0.880	0.120	–	–	2.0048	1,673	−240.9
0.880	0.120	–	–	2.0086	1,673	−228.2

(*Continued*)

TABLE 1 (*Continued*)
Oxygen Potential

U	Pu	Am	Np	O/M	Temp. (K)	$\Delta\bar{G}_{O_2}$ (kJ/mol)
0.880	0.120	–	–	2.0119	1,673	−226.3
0.880	0.120	–	–	1.9950	1,773	−417.9
0.880	0.120	–	–	1.9980	1,773	−382.7
0.880	0.120	–	–	1.9952	1,773	−421.4
0.880	0.120	.	–	1.9970	1,773	−412.3
0.880	0.120	–	–	1.9963	1,773	−408.7
0.880	0.120	–	–	1.9974	1,773	−394.0
0.880	0.120	–	–	1.9990	1,773	−384.8
0.880	0.120	–	–	1.9990	1,773	−361.6
0.880	0.120	–	–	1.9986	1,773	−375.7
0.880	0.120	–	–	1.9993	1,773	−343.3
0.880	0.120	–	–	2.0000	1,773	−322.2
0.880	0.120	–	–	2.0000	1,773	−311.6
0.880	0.120	–	–	2.0012	1,773	−282.8
0.880	0.120	–	–	2.0020	1,773	−272.9
0.880	0.120	–	–	2.0121	1,773	−225.8
0.880	0.120	–	–	2.0270	1,773	−214.5
0.880	0.120	–	–	2.0017	1,773	−272.2
0.880	0.120	–	–	1.9941	1,873	−403.2
0.880	0.120	–	–	1.9954	1,873	−394.3
0.880	0.120	–	–	1.9978	1,873	−389.9
0.880	0.120	–	–	1.9988	1,873	−381.7
0.880	0.120	–	–	1.9973	1,873	−374.2
0.880	0.120	–	–	1.9929	1,873	−421.1
0.880	0.120	–	–	1.9982	1,873	−361.6
0.880	0.120	–	–	2.0008	1,873	−308.8
0.880	0.120	–	–	1.9991	1,873	−333.4
0.880	0.120	–	–	1.9996	1,873	−302.1
0.880	0.120	–	–	2.0018	1,873	−262.7
0.880	0.120	–	–	1.9941	1,873	−408.4
0.880	0.120	–	–	2.0053	1,873	−246.4
0.880	0.120	–	–	1.9998	1,873	−315.5
0.880	0.120	–	–	1.9975	1,873	−376.5
0.880	0.120	–	–	2.0018	1,873	−273.9
0.880	0.120	Am	–	2.0190	1,873	−238.9
0.880	0.120	–	–	2.0408	1,873	−201.8
0.880	0.120	–	–	1.9997	1,873	−286.0
0.800	0.200	–	–	2.0000	1,773	−320.8
0.800	0.200	–	–	2.0000	1,773	−317.3
0.800	0.200	–	–	2.0004	1,773	−290.5

(*Continued*)

TABLE 1 (*Continued*)
Oxygen Potential

U	Pu	Am	Np	O/M	Temp. (K)	$\Delta \bar{G}_{O_2}$ (kJ/mol)
0.800	0.200	–	–	1.9992	1,773	−346.8
0.800	0.200	–	–	2.0009	1,773	−263.8
0.800	0.200	–	–	2.0010	1,773	−277.2
0.800	0.200	–	–	1.9984	1,773	−352.4
0.800	0.200	–	–	2.0018	1,773	−247.6
0.800	0.200	–	–	1.9982	1,773	−360.9
0.800	0.200	–	–	1.9969	1,773	−378.5
0.800	0.200	–	–	1.9966	1,773	−376.4
0.800	0.200	–	–	1.9945	1,773	−401.7
0.800	0.200	–	–	1.9938	1,773	−406.6
0.800	0.200	–	–	1.9914	1,773	−421.4
0.800	0.200	–	–	2.0088	1,773	−209.6
0.800	0.200	–	–	1.9903	1,773	−423.5
0.800	0.200	–	–	1.9886	1,773	−434.1
0.800	0.200	–	–	1.9874	1,773	−441.1
0.800	0.200	–	–	1.9872	1,773	−441.1
0.800	0.200	–	–	1.9826	1,773	−451.0
0.800	0.200	–	–	1.9803	1,773	−458.0
0.800	0.200	–	–	1.9795	1,873	−441.6
0.800	0.200	–	–	1.9819	1,873	−438.3
0.800	0.200	–	–	1.9824	1,873	−426.6
0.800	0.200	–	–	1.9852	1,873	−420.6
0.800	0.200	–	–	1.9866	1,873	−412.5
0.800	0.200	–	–	1.9884	1,873	−404.7
0.800	0.200	–	–	1.9917	1,873	−403.1
0.800	0.200	–	–	1.9945	1,873	−384.8
0.800	0.200	–	–	1.9984	1,873	−355.7
0.800	0.200	–	–	1.9987	1,873	−340.9
0.800	0.200	–	–	1.9989	1,873	−337.8
0.800	0.200	–	–	1.9992	1,873	−323.3
0.800	0.200	–	–	1.9994	1,873	−319.2
0.800	0.200	–	–	1.9997	1,873	−304.1
0.800	0.200	–	–	1.9999	1,873	−308.8
0.800	0.200	–	–	2.0019	1,873	−234.5
0.800	0.200	–	–	2.0019	1,873	−233.7
0.800	0.200	–	–	2.0054	1,873	−195.1
0.800	0.200	–	–	2.0000	1,623	−349.5
0.800	0.200	–	–	1.9916	1,623	−466.0
0.800	0.200	–	–	2.0001	1,623	−362.4
0.800	0.200	–	–	1.9911	1,623	−468.0

(*Continued*)

TABLE 1 (*Continued*)
Oxygen Potential

U	Pu	Am	Np	O/M	Temp. (K)	$\Delta \bar{G}_{O_2}$ (kJ/mol)
0.800	0.200	–	–	1.9996	1,623	−365.6
0.800	0.200	–	–	1.9987	1,623	−389.4
0.800	0.200	–	–	1.9976	1,623	−404.9
0.800	0.200	–	–	1.9965	1,623	−421.0
0.800	0.200	–	–	1.9943	1,623	−442.9
0.800	0.200	–	–	1.9932	1,623	−452.5
0.800	0.200	–	–	1.9995	1,623	−363.7
0.800	0.200	–	–	1.9990	1,623	−375.2
0.800	0.200	–	–	1.9984	1,623	−397.1
0.800	0.200	–	–	1.9953	1,623	−437.1
0.800	0.200	–	–	1.9923	1,623	−459.0
0.800	0.200	–	–	1.9995	1,623	−363.0
0.800	0.200	–	–	1.9930	1,573	−467.5
0.800	0.200	–	–	1.9933	1,573	−465.6
0.800	0.200	–	–	1.9951	1,573	−460.0
0.800	0.200	–	–	1.9956	1,573	−452.5
0.800	0.200	–	–	1.9963	1,573	−448.8
0.800	0.200	–	–	1.9966	1,573	−442.5
0.800	0.200	–	–	1.9976	1,573	−426.9
0.800	0.200	–	–	1.9978	1,573	−425.7
0.800	0.200	–	–	1.9982	1,573	−411.3
0.800	0.200	–	–	1.9991	1,573	−408.8
0.800	0.200	–	–	1.9992	1,573	−403.8
0.800	0.200	–	–	1.9995	1,573	−395.7
0.800	0.200	–	–	1.9997	1,573	−386.4
0.800	0.200	–	–	1.9999	1,573	−371.4
0.800	0.200	–	–	1.9999	1,573	−370.8
0.800	0.200	–	–	2.0000	1,573	−369.5
0.800	0.200	–	–	2.0000	1,573	−365.8
0.800	0.200	–	–	2.0000	1,573	−365.1
0.800	0.200	–	–	2.0000	1,573	−365.1
0.800	0.200	–	–	2.0000	1,573	−363.9
0.800	0.200	–	–	2.0000	1,573	−355.2
0.800	0.200	–	–	2.0000	1,573	−354.5
0.800	0.200	–	–	2.0000	1,573	−353.9
0.800	0.200	–	–	2.0000	1,573	−353.3
0.800	0.200	–	–	2.0000	1,573	−352.7
0.800	0.200	–	–	2.0000	1,573	−351.4
0.800	0.200	–	–	2.0000	1,573	−354.5
0.800	0.200	–	–	2.0000	1,523	−371.6

(*Continued*)

TABLE 1 (*Continued*)
Oxygen Potential

U	Pu	Am	Np	O/M	Temp. (K)	$\Delta \bar{G}_{O_2}$ (kJ/mol)
0.800	0.200	–	–	1.9956	1,523	−461.7
0.800	0.200	–	–	2.0000	1,523	−372.2
0.800	0.200	–	–	1.9966	1,523	−453.8
0.800	0.200	–	–	2.0000	1,523	−371.6
0.800	0.200	–	–	2.0000	1,523	−355.9
0.800	0.200	–	–	1.9953	1,523	−468.9
0.800	0.200	–	–	2.0000	1,523	−357.1
0.800	0.200	–	–	1.9991	1,523	−427.8
0.800	0.200	–	–	2.0000	1,523	−362.6
0.800	0.200	–	–	1.9979	1,523	−438.7
0.800	0.200	–	–	2.0000	1,523	−362.0
0.800	0.200	–	–	1.9994	1,523	−412.7
0.800	0.200	–	–	2.0000	1,523	−362.0
0.800	0.200	–	–	1.9997	1,523	−391.6
0.800	0.200	–	–	2.0000	1,523	−361.3
0.800	0.200	–	–	1.9987	1,523	−429.0
0.800	0.200	–	–	2.0000	1,473	−375.0
0.800	0.200	–	–	1.9983	1,473	−460.9
0.800	0.200	–	–	1.9984	1,473	−463.8
0.800	0.200	–	–	2.0000	1,473	−376.2
0.700	0.300	–	–	1.9788	1,773	−463.8
0.700	0.300	–	–	1.9776	1,773	−446.0
0.700	0.300	–	–	1.9786	1,773	−443.2
0.700	0.300	–	–	1.9851	1,773	−445.9
0.700	0.300	–	–	1.9864	1,773	−429.8
0.700	0.300	–	–	1.9801	1,773	−429.1
0.700	0.300	–	–	1.9890	1,773	−417.4
0.700	0.300	–	–	1.9880	1,773	−413.0
0.700	0.300	–	–	1.9931	1,773	−394.0
0.700	0.300	–	–	1.9936	1,773	−394.0
0.700	0.300	–	–	1.9973	1,773	−395.5
0.700	0.300	–	–	1.9943	1,773	−384.1
0.700	0.300	–	–	1.9960	1,773	−375.0
0.700	0.300	–	–	1.9965	1,773	−375.0
0.700	0.300	–	–	1.9978	1,773	−361.6
0.700	0.300	–	–	1.9983	1,773	−361.6
0.700	0.300	–	–	1.9973	1,773	−360.9
0.700	0.300	–	–	1.9980	1,773	−349.1
0.700	0.300	–	–	1.9969	1,773	−347.5
0.700	0.300	–	–	1.9989	1,773	−338.2

(Continued)

TABLE 1 (*Continued*)
Oxygen Potential

U	Pu	Am	Np	O/M	Temp. (K)	$\Delta \bar{G}_{O_2}$ (kJ/mol)
0.700	0.300	–	–	1.9992	1,773	−337.1
0.700	0.300	–	–	1.9988	1,773	−337.0
0.700	0.300	–	–	2.0003	1,773	−289.8
0.700	0.300	–	–	2.0001	1,773	−291.0
0.700	0.300	–	–	2.0007	1,773	−282.1
0.700	0.300	–	–	2.0006	1,773	−272.2
0.700	0.300	–	–	2.0011	1,773	−272.2
0.700	0.300	–	–	2.0012	1,773	−270.3
0.700	0.300	–	–	2.0008	1,773	−266.6
0.700	0.300	–	–	2.0007	1,773	−258.9
0.700	0.300	–	–	2.0012	1,773	−258.9
0.700	0.300	–	–	2.0017	1,773	−237.0
0.700	0.300	–	–	2.0022	1,773	−237.0
0.700	0.300	–	–	2.0012	1,773	−230.0
0.700	0.300	–	–	2.0032	1,773	−226.0
0.700	0.300	–	–	2.0033	1,773	−206.8
0.700	0.300	–	–	2.0045	1,773	−205.6
0.700	0.300	–	–	2.0028	1,773	−205.4
0.700	0.300	–	–	2.0067	1,773	−189.4
0.700	0.300	–	–	2.0094	1,773	−164.0
0.700	0.300	–	–	2.0265	1,773	−155.4
0.700	0.300	–	–	2.0288	1,773	−139.9
0.700	0.300	–	–	1.9637	1,873	−445.6
0.700	0.300	–	–	1.9831	1,873	−429.4
0.700	0.300	–	–	1.9799	1,873	−428.5
0.700	0.300	–	–	1.9877	1,873	−410.9
0.700	0.300	–	–	1.9888	1,873	−402.7
0.700	0.300	–	–	1.9841	1,873	−390.6
0.700	0.300	–	–	1.9913	1,873	−389.3
0.700	0.300	–	–	1.9914	1,873	−378.7
0.700	0.300	–	–	1.9950	1,873	−370.0
0.700	0.300	–	–	1.9923	1,873	−369.0
0.700	0.300	–	–	1.9970	1,873	−343.2
0.700	0.300	–	–	1.9987	1,873	−329.8
0.700	0.300	–	–	1.9983	1,873	−328.9
0.700	0.300	–	–	1.9995	1,873	−324.4
0.700	0.300	–	–	1.9988	1,873	−320.2
0.700	0.300	–	–	1.9990	1,873	−303.8
0.700	0.300	–	–	1.9989	1,873	−300.1
0.700	0.300	–	–	2.0010	1,873	−293.2

(*Continued*)

TABLE 1 (*Continued*)
Oxygen Potential

U	Pu	Am	Np	O/M	Temp. (K)	$\Delta \bar{G}_{O_2}$ (kJ/mol)
0.700	0.300	–	–	1.9993	1,873	−290.4
0.700	0.300	–	–	1.9996	1,873	−290.4
0.700	0.300	–	–	1.9997	1,873	−280.8
0.700	0.300	–	–	2.0006	1,873	−272.4
0.700	0.300	–	–	2.0007	1,873	−265.2
0.700	0.300	–	–	2.0010	1,873	−227.1
0.700	0.300	–	–	2.0026	1,873	−221.9
0.700	0.300	–	–	2.0017	1,873	−216.1
0.700	0.300	–	–	2.0023	1,873	−208.7
0.700	0.300	–	–	2.0052	1,873	−203.3
0.700	0.300	–	–	2.0029	1,873	−199.0
0.700	0.300	–	–	2.0200	1,873	−182.5
0.700	0.300	–	–	2.0000	1,573	−310.3
0.700	0.300	–	–	1.9998	1,573	−353.3
0.700	0.300	–	–	1.9998	1,573	−362.1
0.700	0.300	–	–	1.9996	1,573	−371.4
0.700	0.300	–	–	1.9995	1,573	−378.3
0.700	0.300	–	–	1.9992	1,573	−385.8
0.700	0.300	–	–	1.9988	1,573	−397.0
0.700	0.300	–	–	1.9984	1,573	−402.0
0.700	0.300	–	–	1.9984	1,573	−405.2
0.700	0.300	–	–	1.9984	1,573	−405.8
0.700	0.300	–	–	1.9979	1,573	−409.5
0.700	0.300	–	–	1.9974	1,573	−416.4
0.700	0.300	–	–	1.9962	1,573	−423.9
0.700	0.300	–	–	1.9953	1,573	−430.1
0.700	0.300	–	–	1.9943	1,573	−435.7
0.700	0.300	–	–	1.9936	1,573	−440.7
0.700	0.300	–	–	1.9923	1,573	−447.6
0.700	0.300	–	–	1.9910	1,573	−453.2
0.700	0.300	–	–	1.9900	1,573	−458.2
0.700	0.300	–	–	1.9865	1,573	−472.1
0.700	0.300	–	–	2.0000	1,623	−301.2
0.700	0.300	–	–	2.0000	1,623	−306.4
0.700	0.300	–	–	1.9998	1623	−337.9
0.700	0.300	–	–	1.9998	1,623	−340.5
0.700	0.300	–	–	1.9997	1,623	−341.2
0.700	0.300	–	–	1.9996	1,623	−342.4
0.700	0.300		–	1.9993	1,623	−360.5

(*Continued*)

TABLE 1 (*Continued*)
Oxygen Potential

U	Pu	Am	Np	O/M	Temp. (K)	$\Delta \bar{G}_{O_2}$ (kJ/mol)
0.700	0.300	–	–	1.9992	1,623	−361.1
0.700	0.300	–	–	1.9991	1,623	−374.0
0.700	0.300	–	–	1.9982	1,623	−381.1
0.700	0.300	–	–	1.9975	1,623	−398.5
0.700	0.300	–	–	1.9970	1,623	−399.8
0.700	0.300	–	–	1.9959	1,623	−410.7
0.700	0.300	–	–	1.9962	1,623	−412.0
0.700	0.300	–	–	1.9951	1,623	−419.1
0.700	0.300	–	–	1.9946	1,623	−423.6
0.700	0.300	–	–	1.9948	1,623	−424.9
0.700	0.300	–	–	1.9939	1,623	−426.8
0.700	0.300	–	–	1.9933	1,623	−429.4
0.700	0.300	–	–	1.9931	1,623	−430.7
0.700	0.300	–	–	1.9923	1,623	−434.6
0.700	0.300	–	–	1.9922	1,623	−435.9
0.700	0.300	–	–	1.9920	1,623	−437.1
0.700	0.300	–	–	1.9917	1,623	−439.1
0.700	0.300	–	–	1.9908	1,623	−442.9
0.700	0.300	–	–	1.9897	1,623	−448.7
0.700	0.300	–	–	1.9871	1,623	−455.8
0.700	0.300	–	–	1.9875	1,623	−454.3
0.700	0.300	–	–	1.9833	1,623	−466.1
0.700	0.300	–	–	1.9812	1,623	−471.3
0.700	0.300	–	–	1.9786	1,623	−471.9
0.700	0.300	–	–	2.0000	1,473	−319.5
0.700	0.300	–	–	1.9998	1,473	−359.3
0.700	0.300	–	–	1.9997	1,473	−377.4
0.700	0.300	–	–	1.9995	1,473	−386.8
0.700	0.300	–	–	1.9994	1,473	−393.8
0.700	0.300	–	–	1.9993	1,473	−402.6
0.700	0.300	–	–	1.9990	1,473	−408.4
0.700	0.300	–	–	1.9990	1,473	−410.2
0.700	0.300	–	–	1.9990	1,473	−410.7
0.700	0.300	–	–	1.9989	1,473	−416.6
0.700	0.300	–	–	1.9986	1,473	−420.7
0.700	0.300	–	–	1.9981	1,473	−427.1
0.700	0.300	–	–	1.9980	1,473	−433.0
0.700	0.300	–	–	1.9972	1,473	−442.9
0.700	0.300	–	–	1.9966	1,473	−449.3
0.700	0.300	–	–	1.9963	1,473	−455.2

(Continued)

TABLE 1 (*Continued*)
Oxygen Potential

U	Pu	Am	Np	O/M	Temp. (K)	$\Delta \bar{G}_{O_2}$ (kJ/mol)
0.700	0.300	–	–	1.9958	1,473	−459.9
0.700	0.300	–	–	2.0000	1,273	−371.4
0.700	0.300	–	–	2.0000	1,273	−380.5
0.700	0.300	–	–	2.0000	1,273	−393.6
0.700	0.300	–	–	1.9999	1,273	−409.3
0.700	0.300	–	–	1.9998	1,273	−423.5
0.700	0.300	–	–	1.9997	1,273	−431.0
0.700	0.300	–	–	1.9997	1,273	−431.0
0.700	0.300	–	–	1.9997	1,273	−432.0
0.700	0.300	–	–	1.9997	1,273	−433.1
0.700	0.300	–	–	1.9997	1,273	−437.1
0.700	0.300	–	–	1.9996	1,273	−444.7
0.700	0.300	–	–	1.9997	1,273	−453.3
0.700	0.300	–	–	1.9995	1,273	−460.3
0.000	1.000	–	–	1.9229	1,673	−436.7
0.000	1.000	–	–	1.9356	1,673	−422.1
0.000	1.000	–	–	1.9537	1,673	−414.8
0.000	1.000	–	–	1.9608	1,673	−409.5
0.000	1.000	–	–	1.9616	1,673	−404.2
0.000	1.000	–	–	1.9618	1,673	−399.6
0.000	1.000	–	–	1.9705	1,673	−392.9
0.000	1.000	–	–	1.9723	1,673	−389.6
0.000	1.000	–	–	1.9727	1,673	−388.3
0.000	1.000	–	–	1.9733	1,673	−389.6
0.000	1.000	–	–	1.9733	1,673	−389.6
0.000	1.000	–	–	1.9753	1,673	−378.3
0.000	1.000	–	–	1.9763	1,673	−386.3
0.000	1.000	–	–	1.9815	1,673	−369.0
0.000	1.000	–	–	1.9863	1,673	−353.7
0.000	1.000	–	–	1.9882	1,673	−343.1
0.000	1.000	–	–	1.9886	1,673	−343.1
0.000	1.000	–	–	1.9910	1,673	−330.5
0.000	1.000	–	–	1.9919	1,673	−325.9
0.000	1.000	–	–	1.9944	1,673	−302.6
0.000	1.000	–	–	1.9945	1,673	−300.0
0.000	1.000	–	–	1.9946	1673	−300.0
0.000	1.000	–	–	1.9963	1,673	−290.7
0.000	1.000	–	–	1.9969	1,673	−286.0
0.000	1.000	–	–	1.9974	1,673	−278.0
0.000	1.000	–	–	1.9988	1,673	−245.5

(*Continued*)

TABLE 1 (*Continued*)
Oxygen Potential

U	Pu	Am	Np	O/M	Temp. (K)	$\Delta \bar{G}_{O_2}$ (kJ/mol)
0.000	1.000	–	–	1.9993	1,673	−232.2
0.000	1.000	–	–	1.9994	1,673	−236.2
0.000	1.000	–	–	2.0000	1,673	−61.2
0.000	1.000	–	–	1.8686	1,773	−431.5
0.000	1.000	–	–	1.8868	1,773	−425.3
0.000	1.000	–	–	1.9012	1,773	−416.9
0.000	1.000	–	–	1.9239	1,773	−404.9
0.000	1.000	–	–	1.9256	1,773	−404.9
0.000	1.000	–	–	1.9428	1,773	−391.8
0.000	1.000	–	–	1.9451	1,773	−391.8
0.000	1.000	–	–	1.9481	1,773	−389.8
0.000	1.000	–	–	1.9526	1,773	−385.5
0.000	1.000	–	–	1.9532	1,773	−381.8
0.000	1.000	–	–	1.9576	1,773	−380.5
0.000	1.000	–	–	1.9581	1,773	−377.9
0.000	1.000	–	–	1.9631	1,773	−370.9
0.000	1.000	–	–	1.9636	1,773	−368.1
0.000	1.000	–	–	1.9692	1,773	−363.6
0.000	1.000	–	–	1.9781	1,773	−344.8
0.000	1.000	–	–	1.9801	1,773	−340.5
0.000	1.000	–	–	1.9829	1,773	−333.2
0.000	1.000	–	–	1.9855	1,773	−329.6
0.000	1.000	–	–	1.9860	1,773	−316.8
0.000	1.000	–	–	1.9866	1,773	−320.1
0.000	1.000	–	–	1.9877	1,773	−311.2
0.000	1.000	–	–	1.9878	1,773	−322.2
0.000	1.000	–	–	1.9914	1,773	−288.7
0.000	1.000	–	–	1.9914	1,773	−289.5
0.000	1.000	–	–	1.9931	1,773	−277.7
0.000	1.000	–	–	1.9947	1,773	−271.3
0.000	1.000	–	–	1.9948	1,773	−266.2
0.000	1.000	–	–	1.9951	1,773	−256.0
0.000	1.000	–	–	1.9952	1,773	−255.5
0.000	1.000	–	–	1.9963	1,773	−238.6
0.000	1.000	–	–	1.9971	1,773	−256.7
0.000	1.000	–	–	1.9976	1,773	−212.2
0.000	1.000	–	–	1.9980	1,773	−212.4
0.000	1.000	–	–	1.9990	1,773	−141.5
0.000	1.000	–	–	2.0000	1,773	−81.0
0.000	1.000	–	–	2.0000	1,773	−67.6

(*Continued*)

TABLE 1 (*Continued*)
Oxygen Potential

U	Pu	Am	Np	O/M	Temp. (K)	$\Delta \bar{G}_{O_2}$ (kJ/mol)
0.000	1.000	–	–	2.0000	1,773	−54.3
0.000	1.000	–	–	2.0000	1,773	−99.3
0.000	1.000	–	–	1.8421	1,873	−432.2
0.000	1.000	–	–	1.8608	1,873	−422.6
0.000	1.000	–	–	1.8727	1,873	−415.9
0.000	1.000	–	–	1.9040	1,873	−412.6
0.000	1.000	–	–	1.9012	1,873	−399.5
0.000	1.000	–	–	1.9168	1,873	−383.9
0.000	1.000	–	–	1.9287	1,873	−381.7
0.000	1.000	–	–	1.9288	1,873	−381.7
0.000	1.000	–	–	1.9288	1,873	−380.9
0.000	1.000	–	–	1.9298	1,873	−380.2
0.000	1.000	–	–	1.9318	1,873	−377.2
0.000	1.000	–	–	1.9425	1,873	−369.8
0.000	1.000	–	–	1.9306	1,873	−358.6
0.000	1.000	–	–	1.9508	1,873	−357.9
0.000	1.000	–	–	1.9511	1,873	−355.7
0.000	1.000	–	–	1.9648	1,873	−344.5
0.000	1.000	–	–	1.9618	1,873	−342.3
0.000	1.000	–	–	1.9627	1,873	−338.6
0.000	1.000	–	–	1.9700	1,873	−331.1
0.000	1.000	–	–	1.9709	1,873	−320.7
0.000	1.000	–	–	1.9769	1,873	−318.5
0.000	1.000	–	–	1.9818	1,873	−304.4
0.000	1.000	–	–	1.9814	1,873	−302.1
0.000	1.000	–	–	1.9873	1,873	−279.8
0.000	1.000	–	–	1.9888	1,873	−271.7
0.000	1.000	–	–	1.9893	1,873	−261.2
0.000	1.000	–	–	1.9908	1,873	−260.5
0.000	1.000	–	–	1.9931	1,873	−243.4
0.000	1.000	–	–	1.9940	1,873	−233.0
0.000	1.000	–	–	1.9955	1,873	−213.7
0.000	1.000	–	–	1.9966	1,873	−199.5
0.000	1.000	–	–	1.9968	1,873	−198.8
0.000	1.000	Am	–	2.0000	1,873	−85.6
0.000	1.000	–	–	2.0000	1,873	−59.5
0.000	1.000	–	–	1.9984	1,473	−311.9
0.000	1.000	–	–	2.0000	1,473	−56.8
0.000	1.000	–	–	2.0000	1,473	−39.8
0.000	1.000	–	–	1.9976	1,473	−322.5

(*Continued*)

TABLE 1 (*Continued*)
Oxygen Potential

U	Pu	Am	Np	O/M	Temp. (K)	$\Delta\bar{G}_{O_2}$ (kJ/mol)
0.000	1.000	–	–	2.0000	1,473	−43.3
0.000	1.000	–	–	1.9995	1,473	−263.4
0.000	1.000	–	–	1.9987	1,473	−312.5
0.000	1.000	–	–	1.9946	1,473	−359.9
0.000	1.000	–	–	1.9977	1,473	−321.9
0.000	1.000	-	-	1.9967	1,473	−342.9
0.000	1.000	-	-	1.9954	1,473	−354.0
0.000	1.000	-	-	1.9946	1,473	−361.1
0.516	0.453	0.031	0.000	1.9995	1,773	−270.2
0.516	0.453	0.031	0.000	1.9988	1,773	−288.5
0.516	0.453	0.031	0.000	1.9986	1,773	−299.1
0.516	0.453	0.031	0.000	1.9977	1,773	−311.0
0.516	0.453	0.031	0.000	1.9997	1,773	−220.3
0.516	0.453	0.031	0.000	1.9984	1,773	−308.3
0.516	0.453	0.031	0.000	1.9977	1,773	−327.3
0.516	0.453	0.031	0.000	1.9973	1,773	−327.3
0.516	0.453	0.031	0.000	1.9985	1,773	−280.8
0.516	0.453	0.031	0.000	1.9855	1,773	−356.1
0.516	0.453	0.031	0.000	1.9954	1,773	−317.4
0.516	0.453	0.031	0.000	1.9933	1,773	−337.1
0.516	0.453	0.031	0.000	1.9830	1,773	−367.4
0.516	0.453	0.031	0.000	1.9910	1,773	−347.0
0.516	0.453	0.031	0.000	1.9816	1,773	−369.5
0.516	0.453	0.031	0.000	1.9910	1,773	−353.3
0.516	0.453	0.031	0.000	1.9776	1,773	−380.1
0.516	0.453	0.031	0.000	1.9648	1,773	−396.2
0.516	0.453	0.031	0.000	1.9700	1,773	−389.9
0.516	0.453	0.031	0.000	1.9591	1,773	−403.3
0.516	0.453	0.031	0.000	1.9485	1,773	−413.1
0.516	0.453	0.031	0.000	1.9550	1,773	−410.3
0.516	0.453	0.031	0.000	1.9402	1,773	−423.0
0.516	0.453	0.031	0.000	1.9500	1,773	−408.2
0.516	0.453	0.031	0.000	1.9356	1,773	−423.0
0.516	0.453	0.031	0.000	1.9284	1,773	−430.0
0.516	0.453	0.031	0.000	2.0018	1,773	−207.6
0.516	0.453	0.031	0.000	2.0006	1,773	−239.3
0.516	0.453	0.031	0.000	2.0003	1,773	−263.5
0.516	0.453	0.031	0.000	1.9998	1,773	−288.2
0.516	0.453	0.031	0.000	2.0020	1,773	−219.9

(*Continued*)

TABLE 1 (*Continued*)
Oxygen Potential

U	Pu	Am	Np	O/M	Temp. (K)	$\Delta \bar{G}_{O_2}$ (kJ/mol)
0.516	0.453	0.031	0.000	1.9996	1,773	−252.3
0.516	0.453	0.031	0.000	1.9990	1,773	−276.5
0.516	0.453	0.031	0.000	2.0008	1,773	−230.1
0.516	0.453	0.031	0.000	2.0006	1,773	−234.3
0.516	0.453	0.031	0.000	2.0005	1,773	−240.7
0.516	0.453	0.031	0.000	2.0003	1,873	−243.2
0.516	0.453	0.031	0.000	1.9982	1,873	−279.6
0.516	0.453	0.031	0.000	1.9969	1,873	−298.5
0.516	0.453	0.031	0.000	1.9955	1,873	−308.9
0.516	0.453	0.031	0.000	2.0020	1,873	−191.5
0.516	0.453	0.031	0.000	2.0008	1,873	−206.0
0.516	0.453	0.031	0.000	2.0003	1,873	−229.0
0.516	0.453	0.031	0.000	2.0019	1,873	−188.9
0.516	0.453	0.031	0.000	2.0003	1,873	−221.3
0.516	0.453	0.031	0.000	1.9995	1,873	−258.8
0.516	0.453	0.031	0.000	1.9984	1,873	−280.0
0.516	0.453	0.031	0.000	1.9988	1,873	−268.9
0.516	0.453	0.031	0.000	1.9873	1,873	−328.0
0.516	0.453	0.031	0.000	1.9828	1,873	−340.2
0.516	0.453	0.031	0.000	1.9910	1,873	−314.9
0.516	0.453	0.031	0.000	1.9736	1,873	−356.9
0.516	0.453	0.031	0.000	1.9423	1,873	−392.9
0.516	0.453	0.031	0.000	1.9700	1,873	−362.1
0.516	0.453	0.031	0.000	1.9441	1,873	−377.7
0.516	0.453	0.031	0.000	1.9700	1,873	−361.7
0.516	0.453	0.031	0.000	1.9226	1,873	−415.2
0.516	0.453	0.031	0.000	1.8500	1,873	−451.7
0.516	0.453	0.031	0.000	1.8168	1,873	−470.7
0.516	0.453	0.031	0.000	1.8007	1,873	−482.2
0.516	0.453	0.031	0.000	1.9880	1,473	−442.7
0.516	0.453	0.031	0.000	1.9850	1,473	−450.0
0.516	0.453	0.031	0.000	1.9932	1,473	−428.4
0.516	0.453	0.031	0.000	1.9857	1,473	−450.0
0.516	0.453	0.031	0.000	1.9813	1,473	−457.6
0.516	0.453	0.031	0.000	1.9982	1,473	−368.1
0.516	0.453	0.031	0.000	1.9970	1,473	−402.0
0.516	0.453	0.031	0.000	1.9919	1,473	−436.0
0.516	0.453	0.031	0.000	1.9990	1,473	−359.7
0.516	0.453	0.031	0.000	1.9978	1,473	−392.5
0.516	0.453	0.031	0.000	1.9956	1,473	−420.5

(Continued)

TABLE 1 (*Continued*)
Oxygen Potential

U	Pu	Am	Np	O/M	Temp. (K)	$\Delta \bar{G}_{O_2}$ (kJ/mol)
0.516	0.453	0.031	0.000	1.9999	1,473	−320.3
0.516	0.453	0.031	0.000	1.9993	1,473	−356.2
0.516	0.453	0.031	0.000	1.9989	1,473	−372.3
0.516	0.453	0.031	0.000	2.0006	1,473	−294.0
0.516	0.453	0.031	0.000	1.9997	1,473	−326.7
0.516	0.453	0.031	0.000	1.9993	1,473	−342.2
0.516	0.453	0.031	0.000	2.0002	1,473	−270.0
0.516	0.453	0.031	0.000	2.0000	1,473	−287.8
0.516	0.453	0.031	0.000	1.9992	1,473	−361.5
0.516	0.453	0.031	0.000	1.9989	1,473	−368.2
0.516	0.453	0.031	0.000	1.9990	1,473	−396.0
0.516	0.453	0.031	0.000	1.9518	1,473	−431.9
0.300	0.677	0.022	0.001	1.9400	1,873	−365.4
0.300	0.677	0.022	0.001	1.9233	1,873	−377.0
0.300	0.677	0.022	0.001	1.8983	1,873	−392.4
0.300	0.677	0.022	0.001	1.9400	1,873	−363.2
0.300	0.677	0.022	0.001	1.9088	1,873	−384.8
0.300	0.677	0.022	0.001	1.8746	1,873	−403.9
0.300	0.677	0.022	0.001	1.8354	1,873	−424.2
0.300	0.677	0.022	0.001	1.8206	1,873	−431.3
0.300	0.677	0.022	0.001	1.9980	1,873	−239.1
0.300	0.677	0.022	0.001	1.9920	1,873	−279.6
0.300	0.677	0.022	0.001	1.9749	1,873	−319.6
0.300	0.677	0.022	0.001	1.9850	1,873	−296.1
0.300	0.677	0.022	0.001	1.9618	1873	−343.7
0.300	0.677	0.022	0.001	1.9103	1,873	−383.6
0.300	0.677	0.022	0.001	2.0015	1,873	−173.7
0.300	0.677	0.022	0.001	1.9974	1,873	−244.3
0.300	0.677	0.022	0.001	1.9947	1,873	−268.4
0.300	0.677	0.022	0.001	1.9806	1,873	−310.0
0.300	0.677	0.022	0.001	2.0006	1,873	−177.5
0.300	0.677	0.022	0.001	2.0000	1,873	−196.0
0.300	0.677	0.022	0.001	1.9994	1,873	−219.3
0.300	0.677	0.022	0.001	1.9972	1,873	−252.9
0.300	0.677	0.022	0.001	1.9600	1,773	−372.3
0.300	0.677	0.022	0.001	1.9467	1,773	−382.9
0.300	0.677	0.022	0.001	1.9277	1,773	−397.2
0.300	0.677	0.022	0.001	1.9630	1,773	−368.3
0.300	0.677	0.022	0.001	1.9406	1,773	−389.0
0.300	0.677	0.022	0.001	1.9131	1,773	−407.4

(*Continued*)

TABLE 1 (*Continued*)
Oxygen Potential

U	Pu	Am	Np	O/M	Temp. (K)	$\Delta \bar{G}_{O_2}$ (kJ/mol)
0.300	0.677	0.022	0.001	1.8786	1,773	−426.9
0.300	0.677	0.022	0.001	1.8654	1,773	−433.1
0.300	0.677	0.022	0.001	1.9990	1,773	−253.5
0.300	0.677	0.022	0.001	1.9964	1,773	−290.7
0.300	0.677	0.022	0.001	1.9873	1,773	−329.2
0.300	0.677	0.022	0.001	1.9930	1,773	−305.6
0.300	0.677	0.022	0.001	1.9760	1,773	−351.2
0.300	0.677	0.022	0.001	1.9384	1,773	−388.9
0.300	0.677	0.022	0.001	2.0005	1,773	−187.8
0.300	0.677	0.022	0.001	1.9988	1,773	−256.6
0.300	0.677	0.022	0.001	1.9976	1,773	−279.7
0.300	0.677	0.022	0.001	1.9906	1,773	−319.7
0.300	0.677	0.022	0.001	2.0002	1,773	−193.7
0.300	0.677	0.022	0.001	2.0000	1,773	−211.2
0.300	0.677	0.022	0.001	1.9999	1,773	−226.4
0.300	0.677	0.022	0.001	1.9987	1,773	−266.0
0.300	0.677	0.022	0.001	1.9996	1,673	−267.1
0.300	0.677	0.022	0.001	1.9981	1,673	−302.5
0.300	0.677	0.022	0.001	1.9962	1,673	−320.2
0.300	0.677	0.022	0.001	1.9933	1,673	−338.3
0.300	0.677	0.022	0.001	1.9970	1,673	−308.1
0.300	0.677	0.022	0.001	1.9885	1,673	−357.7
0.300	0.677	0.022	0.001	1.9794	1,673	−376.2
0.300	0.677	0.022	0.001	1.9712	1,673	−387.5
0.300	0.677	0.022	0.001	1.9581	1,673	−401.5
0.300	0.677	0.022	0.001	1.9890	1,673	−353.7
0.300	0.677	0.022	0.001	1.9803	1,673	−373.8
0.300	0.677	0.022	0.001	1.9664	1,673	−393.4
0.300	0.677	0.022	0.001	1.9482	1,673	−411.1
0.300	0.677	0.022	0.001	1.9228	1,673	−429.1
0.300	0.677	0.022	0.001	1.9132	1,673	−435.2
0.300	0.677	0.022	0.001	1.9997	1,673	−206.9
0.300	0.677	0.022	0.001	1.9997	1,673	−251.9
0.300	0.677	0.022	0.001	1.9991	1,673	−286.9
0.300	0.677	0.022	0.001	1.9954	1,673	−329.4
0.300	0.677	0.022	0.001	2.0000	1,673	−211.8
0.300	0.677	0.022	0.001	2.0000	1,673	−233.6
0.300	0.677	0.022	0.001	1.9997	1,673	−243.6
0.300	0.677	0.022	0.001	1.9992	1,673	−279.1
1.000	–	–	–	2.0320	1,673	−237.7

(*Continued*)

TABLE 1 (*Continued*)
Oxygen Potential

U	Pu	Am	Np	O/M	Temp. (K)	$\Delta \bar{G}_{O_2}$ (kJ/mol)
1.000	–	–	–	2.0119	1,673	−264.3
1.000	–	–	–	2.0059	1,673	−279.5
1.000	–	–	–	2.0087	1,673	−270.9
1.000	–	–	–	2.0033	1,673	−296.8
1.000	–	–	–	2.0020	1,673	−309.4
1.000	–	–	–	2.0010	1,673	−330.0
1.000	–	–	–	2.0009	1,673	−338.6
1.000	–	–	–	2.0016	1,673	−315.4
1.000	–	–	–	2.0000	1,673	−351.3
1.000	–	–	–	2.0000	1,673	−361.9
1.000	–	–	–	2.0000	1,673	−375.2
1.000	–	–	–	2.0000	1,773	−399.1
1.000	–	–	–	2.0000	1,773	−385.0
1.000	–	–	–	2.0000	1,773	−370.9
1.000	–	–	–	2.0032	1,773	−306.2
1.000	–	–	–	2.0020	1,773	−340.6
1.000	–	–	–	2.0021	1,773	−333.6
1.000	–	–	–	2.0034	1,773	−314.6
1.000	–	–	–	2.0058	1,773	−294.9
1.000	–	–	–	2.0022	1,773	−323.0
1.000	–	–	–	2.0082	1,773	−282.9
1.000	–	–	–	2.0151	1,773	−265.3
1.000	–	–	–	2.0035	1,773	−305.4
1.000	–	–	–	2.0194	1,773	−253.3
1.000	–	–	–	2.0383	1,773	−227.3
1.000	–	–	–	2.0600	1,873	−208.0
1.000	–	–	–	2.0297	1,873	−234.0
1.000	–	–	–	2.0154	1,873	−255.5
1.000	–	–	–	2.0375	1,873	−228.0
1.000	–	–	–	2.0097	1,873	−268.9
1.000	–	–	–	2.0059	1,873	−289.0
1.000	–	–	–	2.0041	1,873	−303.1
1.000	–	–	–	2.0068	1,873	−289.7
1.000	–	–	–	2.0032	1,873	−312.0
1.000	Pu	–	–	2.0020	1,873	−326.2
1.000	–	–	–	2.0012	1,873	−335.8
1.000	–	–	–	2.0002	1,873	−356.7
1.000	–	–	–	2.0020	1,873	−335.8
0.000	0.928	0.072	0.000	2.0000	1,473	−19.3
0.000	0.928	0.072	0.000	2.0000	1,473	−19.3

(*Continued*)

TABLE 1 (*Continued*)
Oxygen Potential

U	Pu	Am	Np	O/M	Temp. (K)	$\Delta \bar{G}_{O_2}$ (kJ/mol)
0.000	0.928	0.072	0.000	1.9994	1,473	−62.5
0.000	0.928	0.072	0.000	1.9952	1,473	−121.0
0.000	0.928	0.072	0.000	1.9780	1,473	−258.8
0.000	0.928	0.072	0.000	1.9725	1,473	−252.3
0.000	0.928	0.072	0.000	1.9720	1,473	−256.4
0.000	0.928	0.072	0.000	1.9635	1,473	−294.7
0.000	0.928	0.072	0.000	1.9610	1,473	−308.3
0.000	0.928	0.072	0.000	1.9610	1,473	−309.8
0.000	0.928	0.072	0.000	1.9580	1,473	−335.3
0.000	0.928	0.072	0.000	1.9572	1,473	−351.3
0.000	0.928	0.072	0.000	1.9560	1,473	−357.9
0.000	0.928	0.072	0.000	1.9545	1,473	−388.8
0.000	0.928	0.072	0.000	1.9545	1,473	−391.3
0.000	0.928	0.072	0.000	1.9503	1,473	−399.4
0.000	0.928	0.072	0.000	1.9489	1,473	−418.0
0.000	0.928	0.072	0.000	1.9481	1,473	−412.0
0.000	0.928	0.072	0.000	2.0000	1,673	−23.6
0.000	0.928	0.072	0.000	1.9963	1,673	−69.7
0.000	0.928	0.072	0.000	1.9853	1,673	−140.8
0.000	0.928	0.072	0.000	1.9670	1,673	−226.0
0.000	0.928	0.072	0.000	1.9670	1,673	−227.8
0.000	0.928	0.072	0.000	1.9660	1,673	−230.9
0.000	0.928	0.072	0.000	1.9572	1,673	−297.9
0.000	0.928	0.072	0.000	1.9553	1,673	−315.2
0.000	0.928	0.072	0.000	1.9530	1,673	−323.2
0.000	0.928	0.072	0.000	1.9503	1,673	−338.7
0.000	0.928	0.072	0.000	1.9500	1,673	−331.6
0.000	0.928	0.072	0.000	1.9487	1,673	−346.6
0.000	0.928	0.072	0.000	1.9454	1,673	−349.7
0.000	0.928	0.072	0.000	1.9343	1,673	−385.0
0.000	0.928	0.072	0.000	1.9330	1,673	−387.2
0.000	0.928	0.072	0.000	1.9268	1,673	−405.0
0.000	0.928	0.072	0.000	1.9148	1,673	−419.2
0.000	0.928	0.072	0.000	1.9027	1,673	−436.1
0.000	0.928	0.072	0.000	2.0000	1,773	−23.3
0.000	0.928	0.072	0.000	2.0000	1,773	−18.7
0.000	0.928	0.072	0.000	2.0000	1,773	−23.3
0.000	0.928	0.072	0.000	1.9948	1,773	−64.5
0.000	0.928	0.072	0.000	1.9881	1,773	−120.0
0.000	0.928	0.072	0.000	1.9871	1,773	−120.0

(*Continued*)

TABLE 1 (*Continued*)
Oxygen Potential

U	Pu	Am	Np	O/M	Temp. (K)	$\Delta \bar{G}_{O_2}$ (kJ/mol)
0.000	0.928	0.072	0.000	1.9810	1,773	−136.9
0.000	0.928	0.072	0.000	1.9805	1,773	−138.6
0.000	0.928	0.072	0.000	1.9793	1,773	−162.5
0.000	0.928	0.072	0.000	1.9788	1,773	−174.9
0.000	0.928	0.072	0.000	1.9771	1,773	−207.0
0.000	0.928	0.072	0.000	1.9750	1,773	−221.4
0.000	0.928	0.072	0.000	1.9670	1,773	−282.3
0.000	0.928	0.072	0.000	1.9587	1,773	−309.4
0.000	0.928	0.072	0.000	1.9506	1,773	−328.8
0.000	0.928	0.072	0.000	1.9450	1,773	−343.5
0.000	0.928	0.072	0.000	1.9410	1,773	−355.5
0.000	0.928	0.072	0.000	1.9350	1,773	−366.4
0.000	0.928	0.072	0.000	1.9158	1,773	−385.1
0.000	0.928	0.072	0.000	1.9003	1,773	−403.0
0.000	0.928	0.072	0.000	1.8817	1,773	−414.6
0.000	0.928	0.072	0.000	2.0000	1,873	−24.6
0.000	0.928	0.072	0.000	2.0000	1,873	−24.6
0.000	0.928	0.072	0.000	1.9930	1,873	−69.3
0.000	0.928	0.072	0.000	1.9817	1,873	−145.2
0.000	0.928	0.072	0.000	1.9810	1,873	−134.5
0.000	0.928	0.072	0.000	1.9805	1,873	−136.9
0.000	0.928	0.072	0.000	1.9803	1,873	−141.2
0.000	0.928	0.072	0.000	1.9802	1,873	−179.0
0.000	0.928	0.072	0.000	1.9785	1,873	−188.2
0.000	0.928	0.072	0.000	1.9785	1,873	−189.3
0.000	0.928	0.072	0.000	1.9710	1,873	−239.1
0.000	0.928	0.072	0.000	1.9670	1,873	−261.8
0.000	0.928	0.072	0.000	1.9613	1,873	−281.2
0.000	0.928	0.072	0.000	1.9542	1,873	−302.0
0.000	0.928	0.072	0.000	1.9530	1,873	−304.2
0.000	0.928	0.072	0.000	1.9485	1,873	−322.5
0.000	0.928	0.072	0.000	1.9235	1,873	−351.2
0.000	0.928	0.072	0.000	1.9201	1,873	−359.8
0.000	0.928	0.072	0.000	1.9160	1,873	−360.1
0.000	0.928	0.072	0.000	1.9100	1,873	−366.0
0.000	0.928	0.072	0.000	1.9015	1,873	−374.4
0.000	0.928	0.072	0.000	1.8970	1,873	−377.3
0.000	0.928	0.072	0.000	1.8927	1,873	−380.4
0.000	0.928	0.072	0.000	1.8829	1,873	−388.3
0.000	0.928	0.072	0.000	1.8637	1,873	−404.5
0.000	0.928	0.072	0.000	1.8558	1,873	−415.5

TABLE 2
Melting Temperature

U	Pu	Am	O/M Ratio	Solidus (K)	Liquidus (K)
1.000	0.000	0.000	2.000	3,140	3,145
1.000	0.000	0.000	2.000	3,111	3,130
1.000	0.000	0.000	2.000	3,134	3,173
0.879	0.118	0.003	2.000	3,077	3,117
0.879	0.118	0.003	1.989	3,093	3,135
0.879	0.118	0.003	1.983	3,084	3,105
0.879	0.118	0.003	1.975	3,085	3,107
0.879	0.118	0.003	1.971	3,100	3,124
0.797	0.199	0.004	2.000	3,052	3,090
0.797	0.199	0.004	1.982	3,059	3,089
0.797	0.199	0.004	1.967	3,066	3,079
0.797	0.199	0.004	1.954	3,074	3,109
0.797	0.199	0.004	1.950	3,079	3,097
0.797	0.199	0.004	1.942	3,092	3,118
0.696	0.298	0.006	2.000	3,030	3,074
0.596	0.397	0.007	2.000	2,997	3,029
0.596	0.397	0.007	2.000	3,009	3,020
0.596	0.397	0.007	1.972	3,035	3,071
0.596	0.397	0.007	1.959	3,025	3,037
0.596	0.397	0.007	1.925	3,073	3,102
0.585	0.396	0.019	2.000	3,000	3,052
0.585	0.396	0.019	2.000	3,006	3,043
0.585	0.396	0.019	1.961	3,021	3,043
0.584	0.383	0.033	2.000	2,988	3,044
0.584	0.383	0.033	2.000	2,998	3,050
0.584	0.383	0.033	2.000	3,010	3,039
0.584	0.383	0.033	1.934	3,067	3,108
0.537	0.428	0.035	2.000	2,992	3,025
0.514	0.463	0.024	2.000	2,971	2,998
0.514	0.463	0.024	1.976	3,010	3,079
0.514	0.463	0.024	1.957	3,009	
0.514	0.463	0.024	1.939	3,020	3,063
0.377	0.600	0.023	2.000	2,940	
0.000	0.979	0.021	1.760	3,031	3,077

TABLE 3
Lattice Parameter

U	Pu	Np	Am	O/M	Lattice Parameter (Å)
0.703	0.29	0	0.007	2.000	5.4490
0.703	0.29	0	0.007	2.000	5.4482
0.703	0.29	0	0.007	2.000	5.4485
0.703	0.29	0	0.007	2.000	5.4490
0.703	0.29	0	0.007	2.000	5.4479
0.703	0.29	0	0.007	1.997	5.4484
0.703	0.29	0	0.007	1.993	5.4492
0.703	0.29	0	0.007	1.986	5.4496
0.703	0.29	0	0.007	1.98	5.4503
0.695	0.3	0	0.005	2.000	5.4475
0.695	0.3	0	0.005	1.999	5.4480
0.695	0.3	0	0.005	1.987	5.4514
0.695	0.3	0	0.005	1.981	5.4518
0.643	0.29	0.06	0.007	2.000	5.4463
0.643	0.29	0.06	0.007	2.000	5.4463
0.643	0.29	0.06	0.007	2.000	5.4468
0.643	0.29	0.06	0.007	1.999	5.4465
0.643	0.29	0.06	0.007	1.998	5.4471
0.643	0.29	0.06	0.007	1.997	5.4490
0.643	0.29	0.06	0.007	1.980	5.4500
0.583	0.29	0.12	0.007	2.000	5.4442
0.583	0.29	0.12	0.007	2.000	5.4441
0.583	0.29	0.12	0.007	2.000	5.4440
0.583	0.29	0.12	0.007	1.998	5.4451
0.583	0.29	0.12	0.007	1.997	5.4445
0.583	0.29	0.12	0.007	1.989	5.4464
0.583	0.29	0.12	0.007	1.987	5.4477
0.676	0.3	0	0.024	2.000	5.4469
0.66	0.3	0.02	0.02	2.000	5.4461
0.66	0.3	0.02	0.02	2.000	5.4466
0.66	0.3	0.02	0.02	1.983	5.4470
0.66	0.3	0.02	0.02	1.976	5.4491
0.66	0.3	0.02	0.02	1.923	5.4536
0.664	0.3	0.018	0.018	2.004	5.4465
0.664	0.3	0.018	0.018	2.000	5.4464
0.664	0.3	0.018	0.018	2.000	5.4466
0.664	0.3	0.018	0.018	2.000	5.4463
0.664	0.3	0.018	0.018	1.986	5.4503
0.664	0.3	0.018	0.018	1.979	5.4515
0.879	0.118	0	0.003	2.000	5.4622
0.879	0.118	0	0.003	1.989	5.4632

(Continued)

TABLE 3 (*Continued*)
Lattice Parameter

U	Pu	Np	Am	O/M	Lattice Parameter (Å)
0.879	0.118	0	0.003	1.983	5.4652
0.879	0.118	0	0.003	1.975	5.4693
0.879	0.118	0	0.003	1.974	5.4699
0.879	0.118	0	0.003	1.971	5.4696
0.797	0.199	0	0.004	2.000	5.4558
0.797	0.199	0	0.004	1.982	5.4595
0.797	0.199	0	0.004	1.967	5.4626
0.797	0.199	0	0.004	1.950	5.4699
0.797	0.199	0	0.004	1.942	5.4734
0.787	0.198	0	0.015	2.000	5.4538
0.787	0.198	0	0.015	2.000	5.4540
0.787	0.198	0	0.015	1.996	5.4554
0.787	0.198	0	0.015	1.993	5.4559
0.787	0.198	0	0.015	1.988	5.4561
0.787	0.198	0	0.015	1.981	5.4593
0.787	0.198	0	0.015	1.967	5.4637
0.787	0.198	0	0.015	1.961	5.4609
0.787	0.198	0	0.015	1.953	5.4661
0.787	0.198	0	0.015	1.947	5.4707
0.696	0.298	0	0.006	2.000	5.4466
0.596	0.397	0	0.007	2.000	5.4404
0.596	0.397	0	0.007	1.972	5.4449
0.585	0.396	0	0.019	2.000	5.4381
0.585	0.396	0	0.019	1.961	5.4473
0.584	0.383	0	0.033	2.000	5.4384
0.537	0.428	0	0.035	2.000	5.4358
0.513	0.463	0	0.024	2.000	5.4332
0.513	0.463	0	0.024	2.000	5.4339
0.513	0.463	0	0.024	1.976	5.4397
0.513	0.463	0	0.024	1.718	5.5067
0.513	0.463	0	0.024	1.733	5.5032
0	0.979	0	0.021	2.000	5.3960
0	0.936	0	0.064	2.000	5.3944
0	0.928	0	0.072	2.000	5.3975
0.703	0.297	0	0	2.000	5.4482
0.601	0.399	0	0	2.000	5.4402
0.55	0.45	0	0	2.000	5.4366
0.514	0.486	0	0	2.000	5.4341
1	0	0	0	2.000	5.4709
0.8	0.2	0	0	2.000	5.4556
0.8	0.2	0	0	2.000	5.4556

(Continued)

TABLE 3 (*Continued*)
Lattice Parameter

U	Pu	Np	Am	O/M	Lattice Parameter (Å)
0.8	0.2	0	0	2.000	5.4556
0.8	0.2	0	0	2.000	5.4556
0.8	0.2	0	0	2.000	5.45553
0.8	0.2	0	0	2.000	5.4551
0.8	0.2	0	0	1.985	5.4587
0.8	0.2	0	0	1.975	5.4617
0.8	0.2	0	0	1.963	5.4638
0.7	0.3	0	0	2.00	5.4479
0.515	0.485	0	0	2.0000	5.4350
0	1	0	0	2.0000	5.3975
0.515	0.485	0	0	2.001	5.4339
0.515	0.485	0	0	2.000	5.4355
0.25	0.75	0	0	2.000	5.4142
1	0	0	0	2.000	5.4710

TABLE 4
Sound Speeds and Mechanical Properties

Pu/M Ratio	O/M Ratio	Fract. Porosity	Longitudinal Wave Speed (m/s)	Transverse Wave Speed (m/s)
0.2	2.00	0.1439	4,380.7	2,407.0
0.2	2.000	0.0986	4,660.6	2,523.2
0.2	2.000	0.0842	4,817.4	2,561.1
0.2	2.000	0.0724	4,835.3	2,567.6
0.2	2.000	0.0715	4,838.2	2,557.3
0.2	2.000	0.0543	5,051.8	2,650.0
0.2	1.985	0.0833	4,794.6	2,538.0
0.2	1.975	0.0679	4,801.0	2,496.4
0.2	1.963	0.0693	4,764.8	2,492.8
0	2.000	0.0539	4,966.5	2,621.9
0.3	2.000	0.0568	5,066.8	2,660.3
0.485	2.000	0.0357	5,189.0	2,708.0
0.75	2.000	0.1419	4,498.0	2,491.0
1	2.000	0.0617	5,094.0	2,713.0

TABLE 5
Thermal Expansion

No.	1	2	3	4	5	6	7	8	9	10	11
U	1.00	0.00	0.00	0.00	0.00	0.00	0.70	0.70	0.70	0.70	0.52
Pu	0.00	1.00	1.00	1.00	1.00	1.00	0.30	0.30	0.30	0.30	0.48
O/M	2.000	2.000	2.000	1.980	1.950	1.920	2.000	1.990	1.980	1.970	2.000
b0	-2.88E-03	-2.49E-03	-2.84E-03	-2.88E-03	-2.96E-03	-3.08E-03	-2.87E-03	-2.92E-03	-2.95E-03	-2.99E03	-2.89E-03
b1	9.50E-06	8.31E-06	9.30E-06	9.45E-06	9.73E-06	10.0E-06	9.44E-06	9.60E-06	9.72E-06	9.85E-06	9.40E-06
b2	2.10E-10	9.78E-10	4.00E-10	3.85E-10	3.80E-10	3.80E-10	2.90E-10	2.65E-10	2.55E-10	2.50E-10	3.30E-10
b3	4.40E-13	1.79E-13	3.00E-13	3.20E-13	3.30E-13	3.40E-13	4.00E-13	4.20E-13	4.35E-13	4.50E-13	3.70E-13
Temp. (K)						LTE					
300	7.80E-07	9.37E-05	-5.90E-06	-1.71E-06	2.11E-06	-3.66E-05	-1.10E-06	-4.81E-06	6.95E-07	-3.50E-07	-3.10E-07
325	2.45E-04	3.18E-04	2.35E-04	2.43E-04	2.54E-04	2.22E-04	2.42E-04	2.42E-04	2.51E-04	2.53E-04	2.43E-04
350	4.90E-04	5.44E-04	4.77E-04	4.88E-04	5.06E-04	4.81E-04	4.87E-04	4.90E-04	5.02E-04	5.07E-04	4.86E-04
375	7.35E-04	7.71E-04	7.20E-04	7.35E-04	7.60E-04	7.41E-04	7.32E-04	7.39E-04	7.54E-04	7.63E-04	7.31E-04
400	9.82E-04	1.00E-03	9.63E-04	9.82E-04	1.01E-03	1.00E-03	9.78E-04	9.89E-04	1.01E-03	1.02E-03	9.76E-04
425	1.23E-03	1.23E-03	1.21E-03	1.23E-03	1.27E-03	1.26E-03	1.23E-03	1.24E-03	1.26E-03	1.28E-03	1.22E-03
450	1.48E-03	1.46E-03	1.45E-03	1.48E-03	1.53E-03	1.53E-03	1.47E-03	1.49E-03	1.52E-03	1.53E-03	1.47E-03
475	1.73E-03	1.70E-03	1.70E-03	1.73E-03	1.78E-03	1.79E-03	1.72E-03	1.74E-03	1.77E-03	1.79E-03	1.72E-03
500	1.98E-03	1.93E-03	1.95E-03	1.98E-03	2.04E-03	2.06E-03	1.97E-03	2.00E-03	2.03E-03	2.05E-03	1.97E-03
525	2.23E-03	2.17E-03	2.20E-03	2.23E-03	2.30E-03	2.32E-03	2.22E-03	2.25E-03	2.29E-03	2.32E-03	2.22E-03
550	2.48E-03	2.40E-03	2.45E-03	2.49E-03	2.56E-03	2.59E-03	2.48E-03	2.51E-03	2.55E-03	2.58E-03	2.47E-03
575	2.74E-03	2.64E-03	2.70E-03	2.74E-03	2.82E-03	2.86E-03	2.73E-03	2.77E-03	2.81E-03	2.84E-03	2.72E-03
600	2.99E-03	2.89E-03	2.95E-03	3.00E-03	3.09E-03	3.13E-03	2.98E-03	3.03E-03	3.07E-03	3.11E-03	2.98E-03
625	3.25E-03	3.13E-03	3.20E-03	3.25E-03	3.35E-03	3.40E-03	3.24E-03	3.29E-03	3.33E-03	3.37E-03	3.23E-03
650	3.50E-03	3.37E-03	3.46E-03	3.51E-03	3.62E-03	3.67E-03	3.50E-03	3.55E-03	3.60E-03	3.64E-03	3.49E-03

(Continued)

TABLE 5 (*Continued*)
Thermal Expansion

No.	1	2	3	4	5	6	7	8	9	10	11
675	3.76E-03	3.62E-03	3.71E-03	3.77E-03	3.88E-03	3.95E-03	3.76E-03	3.81E-03	3.86E-03	3.91E-03	3.75E-03
700	4.02E-03	3.87E-03	3.97E-03	4.03E-03	4.15E-03	4.22E-03	4.02E-03	4.07E-03	4.13E-03	4.18E-03	4.01E-03
725	4.29E-03	4.12E-03	4.23E-03	4.30E-03	4.42E-03	4.50E-03	4.28E-03	4.34E-03	4.40E-03	4.45E-03	4.27E-03
750	4.55E-03	4.37E-03	4.49E-03	4.56E-03	4.69E-03	4.78E-03	4.54E-03	4.61E-03	4.67E-03	4.73E-03	4.53E-03
775	4.81E-03	4.62E-03	4.75E-03	4.82E-03	4.96E-03	5.06E-03	4.81E-03	4.87E-03	4.94E-03	5.00E-03	4.80E-03
800	5.08E-03	4.87E-03	5.01E-03	5.09E-03	5.24E-03	5.34E-03	5.07E-03	5.14E-03	5.21E-03	5.28E-03	5.06E-03
825	5.35E-03	5.13E-03	5.27E-03	5.36E-03	5.51E-03	5.62E-03	5.34E-03	5.42E-03	5.49E-03	5.56E-03	5.33E-03
850	5.62E-03	5.39E-03	5.54E-03	5.63E-03	5.79E-03	5.90E-03	5.61E-03	5.69E-03	5.76E-03	5.84E-03	5.60E-03
875	5.89E-03	5.65E-03	5.80E-03	5.90E-03	6.07E-03	6.19E-03	5.88E-03	5.96E-03	6.04E-03	6.12E-03	5.87E-03
900	6.16E-03	5.91E-03	6.07E-03	6.17E-03	6.35E-03	6.48E-03	6.15E-03	6.24E-03	6.32E-03	6.41E-03	6.14E-03
925	6.44E-03	6.17E-03	6.34E-03	6.44E-03	6.63E-03	6.76E-03	6.43E-03	6.52E-03	6.60E-03	6.69E-03	6.41E-03
950	6.71E-03	6.44E-03	6.61E-03	6.72E-03	6.91E-03	7.05E-03	6.70E-03	6.80E-03	6.89E-03	6.98E-03	6.69E-03
975	6.99E-03	6.71E-03	6.89E-03	7.00E-03	7.19E-03	7.35E-03	6.98E-03	7.08E-03	7.17E-03	7.27E-03	6.96E-03
1,000	7.27E-03	6.98E-03	7.16E-03	7.28E-03	7.48E-03	7.64E-03	7.26E-03	7.37E-03	7.46E-03	7.56E-03	7.24E-03
1,025	7.55E-03	7.25E-03	7.44E-03	7.56E-03	7.77E-03	7.94E-03	7.54E-03	7.65E-03	7.75E-03	7.85E-03	7.52E-03
1,050	7.84E-03	7.52E-03	7.71E-03	7.84E-03	8.06E-03	8.23E-03	7.82E-03	7.94E-03	8.04E-03	8.15E-03	7.80E-03
1,075	8.12E-03	7.80E-03	7.99E-03	8.12E-03	8.35E-03	8.53E-03	8.11E-03	8.23E-03	8.33E-03	8.45E-03	8.09E-03
1,100	8.41E-03	8.07E-03	8.27E-03	8.41E-03	8.64E-03	8.83E-03	8.40E-03	8.52E-03	8.63E-03	8.75E-03	8.37E-03
1,125	8.70E-03	8.35E-03	8.56E-03	8.69E-03	8.94E-03	9.14E-03	8.69E-03	8.81E-03	8.93E-03	9.05E-03	8.66E-03
1,150	8.99E-03	8.63E-03	8.84E-03	8.98E-03	9.23E-03	9.44E-03	8.98E-03	9.11E-03	9.23E-03	9.35E-03	8.95E-03
1,175	9.29E-03	8.91E-03	9.13E-03	9.27E-03	9.53E-03	9.75E-03	9.27E-03	9.41E-03	9.53E-03	9.66E-03	9.24E-03
1,200	9.58E-03	9.20E-03	9.41E-03	9.57E-03	9.83E-03	1.01E-02	9.57E-03	9.71E-03	9.83E-03	9.97E-03	9.53E-03
1,225	9.88E-03	9.49E-03	9.70E-03	9.86E-03	1.01E-02	1.04E-02	9.86E-03	1.00E-02	1.01E-02	1.03E-02	9.83E-03
1,250	1.02E-02	9.77E-03	1.00E-02	1.02E-02	1.04E-02	1.07E-02	1.02E-02	1.03E-02	1.04E-02	1.06E-02	1.01E-02

(*Continued*)

TABLE 5 (*Continued*)
Thermal Expansion

No.	1	2	3	4	5	6	7	8	9	10	11
1,275	1.05E-02	1.01E-02	1.03E-02	1.05E-02	1.07E-02	1.10E-02	1.05E-02	1.06E-02	1.08E-02	1.09E-02	1.04E-02
1,300	1.08E-02	1.04E-02	1.06E-02	1.08E-02	1.11E-02	1.13E-02	1.08E-02	1.09E-02	1.11E-02	1.12E-02	1.07E-02
1,325	1.11E-02	1.07E-02	1.09E-02	1.11E-02	1.14E-02	1.16E-02	1.11E-02	1.12E-02	1.14E-02	1.15E-02	1.10E-02
1,350	1.14E-02	1.10E-02	1.12E-02	1.14E-02	1.17E-02	1.19E-02	1.14E-02	1.16E-02	1.17E-02	1.19E-02	1.13E-02
1,375	1.17E-02	1.13E-02	1.15E-02	1.17E-02	1.20E-02	1.23E-02	1.17E-02	1.19E-02	1.20E-02	1.22E-02	1.17E-02
1,400	1.20E-02	1.16E-02	1.18E-02	1.20E-02	1.23E-02	1.26E-02	1.20E-02	1.22E-02	1.24E-02	1.25E-02	1.20E-02
1,425	1.24E-02	1.19E-02	1.21E-02	1.23E-02	1.26E-02	1.29E-02	1.23E-02	1.25E-02	1.27E-02	1.29E-02	1.23E-02
1,450	1.27E-02	1.22E-02	1.24E-02	1.26E-02	1.30E-02	1.33E-02	1.26E-02	1.28E-02	1.30E-02	1.32E-02	1.26E-02
1,475	1.30E-02	1.25E-02	1.27E-02	1.29E-02	1.33E-02	1.36E-02	1.30E-02	1.32E-02	1.33E-02	1.35E-02	1.29E-02
1,500	1.33E-02	1.28E-02	1.30E-02	1.32E-02	1.36E-02	1.39E-02	1.33E-02	1.35E-02	1.37E-02	1.39E-02	1.32E-02
1,525	1.37E-02	1.31E-02	1.33E-02	1.36E-02	1.39E-02	1.43E-02	1.36E-02	1.38E-02	1.40E-02	1.42E-02	1.36E-02
1,550	1.40E-02	1.34E-02	1.37E-02	1.39E-02	1.43E-02	1.46E-02	1.39E-02	1.42E-02	1.43E-02	1.46E-02	1.39E-02
1,575	1.43E-02	1.37E-02	1.40E-02	1.42E-02	1.46E-02	1.49E-02	1.43E-02	1.45E-02	1.47E-02	1.49E-02	1.42E-02
1,600	1.47E-02	1.40E-02	1.43E-02	1.45E-02	1.49E-02	1.53E-02	1.46E-02	1.48E-02	1.50E-02	1.53E-02	1.45E-02
1,625	1.50E-02	1.44E-02	1.46E-02	1.49E-02	1.53E-02	1.56E-02	1.50E-02	1.52E-02	1.54E-02	1.56E-02	1.49E-02
1,650	1.53E-02	1.47E-02	1.49E-02	1.52E-02	1.56E-02	1.60E-02	1.53E-02	1.55E-02	1.57E-02	1.60E-02	1.52E-02
1,675	1.57E-02	1.50E-02	1.53E-02	1.55E-02	1.60E-02	1.63E-02	1.56E-02	1.59E-02	1.61E-02	1.63E-02	1.55E-02
1,700	1.60E-02	1.53E-02	1.56E-02	1.59E-02	1.63E-02	1.67E-02	1.60E-02	1.62E-02	1.64E-02	1.67E-02	1.59E-02
1,725	1.64E-02	1.57E-02	1.59E-02	1.62E-02	1.66E-02	1.70E-02	1.63E-02	1.66E-02	1.68E-02	1.71E-02	1.62E-02
1,750	1.67E-02	1.60E-02	1.63E-02	1.66E-02	1.70E-02	1.74E-02	1.67E-02	1.69E-02	1.72E-02	1.74E-02	1.66E-02
1,775	1.71E-02	1.63E-02	1.66E-02	1.69E-02	1.74E-02	1.78E-02	1.70E-02	1.73E-02	1.75E-02	1.78E-02	1.69E-02
1,800	1.75E-02	1.67E-02	1.69E-02	1.72E-02	1.77E-02	1.81E-02	1.74E-02	1.77E-02	1.79E-02	1.82E-02	1.73E-02
1,825	1.78E-02	1.70E-02	1.73E-02	1.76E-02	1.81E-02	1.85E-02	1.78E-02	1.80E-02	1.83E-02	1.86E-02	1.76E-02
1,850	1.82E-02	1.74E-02	1.76E-02	1.79E-02	1.84E-02	1.89E-02	1.81E-02	1.84E-02	1.87E-02	1.89E-02	1.80E-02
1,875	1.86E-02	1.77E-02	1.80E-02	1.83E-02	1.88E-02	1.92E-02	1.85E-02	1.88E-02	1.90E-02	1.93E-02	1.84E-02
1,900	1.89E-02	1.81E-02	1.83E-02	1.87E-02	1.92E-02	1.96E-02	1.89E-02	1.92E-02	1.94E-02	1.97E-02	1.87E-02

TABLE 6
Thermal Diffusivity

U	Pu	Am	Np	O/M	ρ (%TD)	Temp. (K)	k (m²/s)	C_p^a (J/molK)	λ^a (W/mK)
0.693	0.300	0.007		2.000	0.915	892.0	1.2E-06	86.8	3.86
0.693	0.300	0.007		2.000	0.915	976.0	1.1E-06	88.0	3.47
0.693	0.300	0.007	-	2.000	0.915	1,076.0	9.5E-07	89.1	3.18
0.693	0.300	0.007	-	2.000	0.915	1,177.0	8.6E-07	90.4	2.93
0.693	0.300	0.007	-	2.000	0.915	894.0	1.2E-06	86.8	3.87
0.693	0.300	0.007	-	2.000	0.915	1,285.0	8.0E-07	92.8	2.69
0.693	0.300	0.007	-	2.000	0.915	1,370.0	7.4E-07	95.1	2.56
0.693	0.300	0.007	-	2.000	0.915	1,475.0	6.9E-07	99.6	2.49
0.693	0.300	0.007	-	2.000	0.915	875.0	1.2E-06	86.7	3.90
0.693	0.300	0.007	-	2.000	0.915	860.0	1.3E-06	86.5	4.00
0.693	0.300	0.007	-	2.000	0.915	1,576.0	6.8E-07	106.4	2.59
0.693	0.300	0.007	-	2.000	0.915	1,575.0	6.5E-07	106.4	2.47
0.693	0.300	0.007	-	2.000	0.915	1,675.0	6.1E-07	116.3	2.56
0.693	0.300	0.007	-	2.000	0.915	1,775.0	5.8E-07	129.9	2.70
0.670	0.300	0.030	-	2.000	0.929	891.0	1.1E-06	86.8	3.65
0.670	0.300	0.030	-	2.000	0.929	973.0	1.0E-06	88.0	3.34
0.670	0.300	0.030	-	2.000	0.929	1,075.0	9.3E-07	89.3	3.09
0.670	0.300	0.030	-	2.000	0.929	1,175.0	8.5E-07	90.6	2.88
0.670	0.300	0.030	-	2.000	0.929	1,280.0	7.9E-07	92.5	2.71
0.670	0.300	0.030	-	2.000	0.929	1,373.0	7.4E-07	95.0	2.59
0.670	0.300	0.030	-	2.000	0.929	1,475.0	7.0E-07	99.4	2.55
0.670	0.300	0.030	-	2.000	0.929	1,573.0	6.4E-07	106.0	2.48
0.670	0.300	0.030	-	2.000	0.929	1,674.0	6.1E-07	115.9	2.58
0.670	0.300	0.030	-	2.000	0.929	1,774.0	5.8E-07	129.6	2.75
0.680	0.300	0.020	-	2.000	0.936	872.0	1.2E-06	86.6	3.86
0.680	0.300	0.020	-	2.000	0.936	975.0	1.0E-06	88.1	3.46
0.680	0.300	0.020	-	2.000	0.936	1,074.0	9.5E-07	89.4	3.18
0.680	0.300	0.020	-	2.000	0.936	1,174.0	8.7E-07	90.7	2.94
0.680	0.300	0.020	-	2.000	0.936	1,279.0	8.1E-07	92.5	2.79
0.680	0.300	0.020	-	2.000	0.936	1,372.0	7.6E-07	95.0	2.68
0.680	0.300	0.020	-	2.000	0.936	1,474.0	7.2E-07	99.4	2.66
0.680	0.300	0.020	-	2.000	0.936	1,573.0	6.7E-07	106.1	2.61
0.680	0.300	0.020	-	2.000	0.936	1,674.0	6.4E-07	116.0	2.72
0.680	0.300	0.020	-	2.000	0.936	1,774.0	6.1E-07	129.6	2.91
0.693	0.300	0.007	-	2.000	0.912	892.0	1.2E-06	87.0	4.00
0.693	0.300	0.007	-	2.000	0.912	972.0	1.1E-06	88.2	3.46
0.693	0.300	0.007	Np	2.000	0.912	1,071.0	9.5E-07	89.5	3.12
0.693	0.300	0.007	-	2.000	0.912	1,167.0	8.8E-07	90.7	2.90
0.693	0.300	0.007	-	2.000	0.912	1,271.0	8.1E-07	92.5	2.71
0.693	0.300	0.007	-	2.000	0.912	1,364.0	7.6E-07	94.9	2.62
0.693	0.300	0.007	-	2.000	0.912	1,465.0	7.1E-07	99.1	2.54
0.693	0.300	0.007	-	2.000	0.912	1,563.0	6.5E-07	105.4	2.46

(Continued)

TABLE 6 (*Continued*)
Thermal Diffusivity

U	Pu	Am	Np	O/M	ρ (%TD)	Temp. (K)	k (m²/s)	$C_p{}^a$ (J/molK)	λ^a (W/mK)
0.693	0.300	0.007	-	2.000	0.912	1,665.0	6.2E-07	115.1	2.53
0.693	0.300	0.007	-	2.000	0.912	1,765.0	5.9E-07	128.3	2.70
0.678	0.300	0.022	-	2.000	0.948	872.0	1.1E-06	86.6	3.79
0.678	0.300	0.022	-	2.000	0.948	974.0	1.0E-06	88.1	3.48
0.678	0.300	0.022	-	2.000	0.948	1,076.0	9.1E-07	89.4	3.09
0.678	0.300	0.022	-	2.000	0.948	1,175.0	8.5E-07	90.7	2.92
0.678	0.300	0.022	-	2.000	0.948	1,279.0	8.0E-07	92.5	2.80
0.678	0.300	0.022	-	2.000	0.948	1,372.0	7.4E-07	95.0	2.65
0.678	0.300	0.022	-	2.000	0.948	1,473.0	7.2E-07	99.4	2.68
0.678	0.300	0.022	-	2.000	0.948	1,571.0	6.4E-07	105.9	2.52
0.678	0.300	0.022	-	2.000	0.948	1,673.0	6.1E-07	115.9	2.62
0.678	0.300	0.022	-	2.000	0.948	1,773.0	5.9E-07	129.4	2.82
0.678	0.300	0.022	-	2.000	0.952	917.0	1.1E-06	87.3	3.82
0.678	0.300	0.022	-	2.000	0.952	979.0	1.0E-06	88.2	3.52
0.678	0.300	0.022	-	2.000	0.952	1,073.0	9.2E-07	89.4	3.15
0.678	0.300	0.022	-	2.000	0.952	1,172.0	8.8E-07	90.7	3.05
0.678	0.300	0.022	-	2.000	0.952	1,277.0	8.1E-07	92.5	2.84
0.678	0.300	0.022	-	2.000	0.952	1,369.0	7.5E-07	94.9	2.71
0.678	0.300	0.022	-	2.000	0.952	1,471.0	6.9E-07	99.3	2.59
0.678	0.300	0.022	-	2.000	0.952	1,569.0	6.7E-07	105.8	2.66
0.678	0.300	0.022	-	2.000	0.952	1,670.0	6.4E-07	115.6	2.75
0.678	0.300	0.022	-	2.000	0.952	1,770.0	6.1E-07	129.0	2.94
0.678	0.300	0.022	-	1.979	0.942	918.0	1.0E-06	86.7	3.42
0.678	0.300	0.022	-	1.979	0.942	985.0	9.3E-07	87.6	3.09
0.678	0.300	0.022	-	1.979	0.942	1,074.0	8.7E-07	88.7	2.92
0.678	0.300	0.022	-	1.979	0.942	1,171.0	8.1E-07	90.0	2.73
0.678	0.300	0.022	-	1.979	0.942	1,276.0	7.6E-07	91.8	2.61
0.678	0.300	0.022	-	1.979	0.942	1,369.0	7.3E-07	94.3	2.57
0.678	0.300	0.022	-	1.979	0.942	1,471.0	6.9E-07	98.5	2.51
0.678	0.300	0.022	-	1.979	0.942	1,569.0	6.6E-07	105.0	2.57
0.678	0.300	0.022	-	1.979	0.942	1,670.0	6.6E-07	114.8	2.79
0.678	0.300	0.022	-	1.979	0.942	1,770.0	6.3E-07	128.3	2.97
0.678	0.300	0.022	-	1.963	0.938	901.0	8.5E-07	85.9	2.77
0.678	0.300	0.022	-	1.963	0.938	972.0	8.0E-07	86.9	2.61
0.678	0.300	0.022	-	1.963	0.938	1,068.0	7.5E-07	88.2	2.46
0.678	0.300	0.022	-	1.963	0.938	1,166.0	6.9E-07	89.4	2.31
0.678	0.300	0.022	-	1.963	0.938	1,271.0	6.6E-07	91.1	2.23
0.678	0.300	0.022	-	1.963	0.938	1,363.0	6.2E-07	93.5	2.16
0.678	0.300	0.022	-	1.963	0.938	1,465.0	6.0E-07	97.7	2.17
0.678	0.300	0.022	-	1.963	0.938	1,563.0	5.7E-07	104.0	2.18
0.678	0.300	0.022	-	1.963	0.938	1,664.0	5.6E-07	113.6	2.32
0.678	0.300	0.022	-	1.963	0.938	1,764.0	5.4E-07	126.8	2.51

(Continued)

TABLE 6 (*Continued*)
Thermal Diffusivity

U	Pu	Am	Np	O/M	ρ (%TD)	Temp. (K)	k (m²/s)	$C_p{}^a$ (J/molK)	$λ^a$ (W/mK)
0.678	0.300	0.022	-	1.946	0.942	874.0	7.6E-07	85.0	2.45
0.678	0.300	0.022	-	1.946	0.942	968.0	7.2E-07	86.4	2.35
0.678	0.300	0.022	-	1.946	0.942	1,068.0	6.8E-07	87.6	2.24
0.678	0.300	0.022	-	1.946	0.942	1,166.0	6.4E-07	88.9	2.13
0.678	0.300	0.022	-	1.946	0.942	1,271.0	6.1E-07	90.6	2.06
0.678	0.300	0.022	-	1.946	0.942	1,364.0	5.9E-07	93.0	2.05
0.678	0.300	0.022	-	1.946	0.942	1,465.0	5.7E-07	97.2	2.04
0.678	0.300	0.022	-	1.946	0.942	1,563.0	5.5E-07	103.4	2.09
0.678	0.300	0.022	-	1.946	0.942	1,665.0	5.4E-07	113.1	2.24
0.678	0.300	0.022	-	1.946	0.942	1,765.0	5.3E-07	126.4	2.47
0.678	0.300	0.022	-	1.922	0.934	875.0	6.5E-07	84.3	2.05
0.678	0.300	0.022	-	1.922	0.934	967.0	6.3E-07	85.6	2.00
0.678	0.300	0.022	-	1.922	0.934	1,067.0	5.9E-07	86.9	1.89
0.678	0.300	0.022	-	1.922	0.934	1,167.0	5.6E-07	88.1	1.83
0.678	0.300	0.022	-	1.922	0.934	1,270.0	5.3E-07	89.8	1.76
0.678	0.300	0.022	-	1.922	0.934	1,363.0	5.2E-07	92.2	1.77
0.678	0.300	0.022	-	1.922	0.934	1,465.0	5.1E-07	96.4	1.78
0.678	0.300	0.022	-	1.922	0.934	1,563.0	4.8E-07	102.6	1.80
0.678	0.300	0.022	-	1.922	0.934	1,664.0	4.7E-07	112.2	1.91
0.678	0.300	0.022	-	1.922	0.934	1,764.0	4.7E-07	125.4	2.11
0.678	0.300	0.022	-	1.915	0.931	882.0	6.2E-07	84.2	1.94
0.678	0.300	0.022	-	1.915	0.931	963.0	6.0E-07	85.4	1.90
0.678	0.300	0.022	-	1.915	0.931	1,064.0	5.6E-07	86.6	1.80
0.678	0.300	0.022	-	1.915	0.931	1,166.0	5.4E-07	87.9	1.75
0.678	0.300	0.022	-	1.915	0.931	1,271.0	5.2E-07	89.6	1.71
0.678	0.300	0.022	-	1.915	0.931	1,363.0	5.0E-07	91.9	1.68
0.678	0.300	0.022	-	1.915	0.931	1,465.0	4.9E-07	96.1	1.71
0.678	0.300	0.022	-	1.915	0.931	1,563.0	4.5E-07	102.4	1.68
0.678	0.300	0.022	-	1.915	0.931	1,664.0	4.4E-07	112.0	1.79
0.678	0.300	0.022	-	1.915	0.931	1,764.0	4.5E-07	125.1	2.00
0.678	0.300	0.022	-	2.000	0.843	883.0	1.1E-06	86.7	3.14
0.678	0.300	0.022	-	2.000	0.843	978.0	9.6E-07	88.2	2.87
0.678	0.300	0.022	-	2.000	0.843	1,072.0	8.7E-07	89.4	2.62
0.678	0.300	0.022	-	2.000	0.843	1,166.0	8.1E-07	90.6	2.46
0.678	0.300	0.022	-	2.000	0.843	1,270.0	7.2E-07	92.3	2.23
0.678	0.300	0.022	-	2.000	0.843	1,363.0	6.9E-07	94.7	2.19
0.678	0.300	0.022	-	2.000	0.843	1,464.0	6.5E-07	98.9	2.14
0.678	0.300	0.022	-	2.000	0.843	1,563.0	6.2E-07	105.3	2.17
0.678	0.300	0.022	-	2.000	0.843	1,665.0	5.9E-07	115.0	2.26
0.678	0.300	0.022	-	2.000	0.843	1,766.0	5.7E-07	128.4	2.42
0.678	0.300	0.022	-	2.000	0.940	872.0	1.1E-06	86.6	3.75
0.678	0.300	0.022	-	2.000	0.940	974.0	1.0E-06	88.1	3.45

(Continued)

TABLE 6 (*Continued*)
Thermal Diffusivity

U	Pu	Am	Np	O/M	ρ (%TD)	Temp. (K)	k (m²/s)	$C_p{}^a$ (J/molK)	λ^a (W/mK)
0.678	0.300	0.022	-	2.000	0.940	1,076.0	9.1E-07	89.4	3.06
0.678	0.300	0.022	-	2.000	0.940	1,175.0	8.5E-07	90.7	2.89
0.678	0.300	0.022	-	2.000	0.940	1,279.0	8.0E-07	92.5	2.77
0.678	0.300	0.022	-	2.000	0.940	1,372.0	7.4E-07	95.0	2.63
0.678	0.300	0.022	-	2.000	0.940	1,473.0	7.2E-07	99.4	2.66
0.678	0.300	0.022	-	2.000	0.940	1,571.0	6.4E-07	105.9	2.50
0.678	0.300	0.022	-	2.000	0.940	1,673.0	6.1E-07	115.9	2.60
0.678	0.300	0.022	-	2.000	0.940	1,773.0	5.9E-07	129.4	2.79
0.678	0.300	0.022	-	2.000	0.896	902.0	1.1E-06	87.0	3.33
0.678	0.300	0.022	-	2.000	0.896	979.0	9.6E-07	88.2	3.07
0.678	0.300	0.022	-	2.000	0.896	1,075.0	8.9E-07	89.4	2.85
0.678	0.300	0.022	-	2.000	0.896	1,172.0	8.0E-07	90.7	2.61
0.678	0.300	0.022	-	2.000	0.896	1,277.0	7.5E-07	92.5	2.47
0.678	0.300	0.022	-	2.000	0.896	1,369.0	7.1E-07	94.9	2.38
0.678	0.300	0.022	-	2.000	0.896	1,471.0	6.7E-07	99.3	2.36
0.678	0.300	0.022	-	2.000	0.896	1,569.0	6.3E-07	105.8	2.36
0.678	0.300	0.022	-	2.000	0.896	1,671.0	6.0E-07	115.6	2.46
0.678	0.300	0.022	-	2.000	0.896	1,770.0	5.9E-07	129.0	2.65
0.678	0.300	0.022	-	2.000	0.920	885.0	1.1E-06	86.8	3.69
0.678	0.300	0.022	-	2.000	0.920	970.0	1.0E-06	88.0	3.40
0.678	0.300	0.022	-	2.000	0.920	1,077.0	9.3E-07	89.4	3.07
0.678	0.300	0.022	-	2.000	0.920	1,174.0	8.7E-07	90.7	2.90
0.678	0.300	0.022	-	2.000	0.920	1,278.0	7.8E-07	92.5	2.63
0.678	0.300	0.022	-	2.000	0.920	1,371.0	7.5E-07	95.0	2.62
0.678	0.300	0.022	-	2.000	0.920	1,473.0	7.1E-07	99.4	2.56
0.678	0.300	0.022	-	2.000	0.920	1,571.0	6.6E-07	105.9	2.52
0.678	0.300	0.022	-	2.000	0.920	1,672.0	6.3E-07	115.8	2.63
0.678	0.300	0.022	-	2.000	0.920	1,772.0	6.1E-07	129.3	2.83
0.678	0.300	0.022	-	2.000	0.926	900.0	1.1E-06	87.0	3.57
0.678	0.300	0.022	-	2.000	0.926	976.0	1.0E-06	88.1	3.31
0.678	0.300	0.022	-	2.000	0.926	1,074.0	9.2E-07	89.4	3.05
0.678	0.300	0.022	-	2.000	0.926	1,174.0	8.3E-07	90.7	2.80
0.678	0.300	0.022	-	2.000	0.926	1,277.0	7.7E-07	92.5	2.62
0.678	0.300	0.022	-	2.000	0.926	1,370.0	7.3E-07	95.0	2.56
0.678	0.300	0.022	-	2.000	0.926	1,472.0	6.8E-07	99.3	2.49
0.678	0.300	0.022	-	2.000	0.926	1,570.0	6.5E-07	105.8	2.50
0.678	0.300	0.022	-	2.000	0.926	1,670.0	6.2E-07	115.6	2.61
0.678	0.300	0.022	-	2.000	0.926	1,770.0	6.0E-07	129.0	2.78
0.678	0.300	0.022	-	2.000	0.952	917.0	1.1E-06	87.3	3.83
0.678	0.300	0.022	-	2.000	0.952	979.0	1.0E-06	88.2	3.53
0.678	0.300	0.022		2.000	0.952	1,073.0	9.2E-07	89.4	3.15
0.678	0.300	0.022	-	2.000	0.952	1,172.0	8.9E-07	90.7	3.06

(Continued)

TABLE 6 (*Continued*)
Thermal Diffusivity

U	Pu	Am	Np	O/M	ρ (%TD)	Temp. (K)	k (m²/s)	C_p^a (J/molK)	$λ^a$ (W/mK)
0.678	0.300	0.022	-	2.000	0.952	1,277.0	8.1E-07	92.5	2.84
0.678	0.300	0.022	-	2.000	0.952	1,369.0	7.6E-07	94.9	2.71
0.678	0.300	0.022	-	2.000	0.952	1,471.0	6.9E-07	99.3	2.59
0.678	0.300	0.022	-	2.000	0.952	1,569.0	6.7E-07	105.8	2.67
0.678	0.300	0.022	-	2.000	0.952	1,670.0	6.4E-07	115.6	2.75
0.678	0.300	0.022	-	2.000	0.952	1,770.0	6.1E-07	129.0	2.94
0.678	0.300	0.022	-	2.000	0.872	902.0	1.1E-06	87.0	3.27
0.678	0.300	0.022	-	2.000	0.872	968.0	9.7E-07	88.0	2.99
0.678	0.300	0.022	-	2.000	0.872	1,069.0	8.6E-07	89.3	2.70
0.678	0.300	0.022	-	2.000	0.872	1,168.0	7.9E-07	90.6	2.49
0.678	0.300	0.022	-	2.000	0.872	1,271.0	7.4E-07	92.3	2.39
0.678	0.300	0.022	-	2.000	0.872	1,363.0	6.8E-07	94.7	2.23
0.678	0.300	0.022	-	2.000	0.872	1,466.0	6.6E-07	99.0	2.25
0.678	0.300	0.022	-	2.000	0.872	1,564.0	6.1E-07	105.4	2.20
0.678	0.300	0.022	-	2.000	0.872	1,666.0	5.8E-07	115.1	2.30
0.678	0.300	0.022	-	2.000	0.872	1,766.0	5.7E-07	128.4	2.49
0.647	0.288	0.007	0.059	2.000	0.936	892.0	1.3E-06	86.7	4.11
0.647	0.288	0.007	0.059	2.000	0.936	969.0	1.1E-06	87.8	3.74
0.647	0.288	0.007	0.059	2.000	0.936	1,069.0	1.0E-06	89.1	3.39
0.647	0.288	0.007	0.059	2.000	0.936	1,165.0	9.2E-07	90.3	3.11
0.647	0.288	0.007	0.059	2.000	0.936	1,269.0	8.4E-07	92.0	2.89
0.647	0.288	0.007	0.059	2.000	0.936	1,361.0	7.9E-07	94.3	2.75
0.647	0.288	0.007	0.059	2.000	0.936	1,464.0	7.4E-07	98.3	2.69
0.647	0.288	0.007	0.059	2.000	0.936	1,562.0	6.7E-07	104.3	2.56
0.647	0.288	0.007	0.059	2.000	0.936	1,664.0	6.4E-07	113.5	2.67
0.647	0.288	0.007	0.059	2.000	0.936	1,764.0	6.1E-07	126.1	2.83
0.583	0.290	0.008	0.119	2.000	0.941	1,008.0	9.9E-07	87.9	3.30
0.583	0.290	0.008	0.119	2.000	0.941	1,104.0	9.0E-07	89.1	3.03
0.583	0.290	0.008	0.119	2.000	0.941	1,185.0	8.3E-07	90.2	2.81
0.583	0.290	0.008	0.119	2.000	0.941	1,291.0	7.7E-07	92.1	2.65
0.583	0.290	0.008	0.119	2.000	0.941	1,386.0	7.1E-07	94.8	2.52
0.583	0.290	0.008	0.119	2.000	0.941	1,489.0	6.6E-07	99.3	2.44
0.583	0.290	0.008	0.119	2.000	0.941	1,588.0	6.2E-07	106.2	2.46
0.583	0.290	0.008	0.119	2.000	0.941	1,691.0	5.9E-07	116.5	2.56
0.583	0.290	0.008	0.119	2.000	0.941	1,793.0	5.7E-07	130.5	2.73
0.583	0.290	0.008	0.119	2.000	0.941	992.0	1.0E-06	87.7	3.35
0.583	0.290	0.008	0.119	2.000	0.941	1,087.0	9.2E-07	88.9	3.08
0.583	0.290	0.008	0.119	2.000	0.941	1,170.0	8.4E-07	90.0	2.85
0.583	0.290	0.008	0.119	2.000	0.941	1,276.0	7.7E-07	91.7	2.66
0.583	0.290	0.008	0.119	2.000	0.941	1,369.0	7.2E-07	94.2	2.55
0.583	0.290	0.008	0.119	2.000	0.941	1,472.0	6.7E-07	98.4	2.47
0.583	0.290	0.008	0.119	2.000	0.941	1,569.0	6.2E-07	104.7	2.43

(*Continued*)

TABLE 6 (*Continued*)
Thermal Diffusivity

U	Pu	Am	Np	O/M	ρ (%TD)	Temp. (K)	k (m²/s)	C_pª (J/molK)	λª (W/mK)
0.583	0.290	0.008	0.119	2.000	0.941	1,670.0	6.0E-07	114.1	2.52
0.583	0.290	0.008	0.119	2.000	0.941	1,769.0	5.7E-07	126.9	2.67
0.647	0.288	0.007	0.059	2.000	0.930	887.0	1.2E-06	86.6	3.93
0.647	0.288	0.007	0.059	2.000	0.930	981.0	1.0E-06	88.0	3.43
0.647	0.288	0.007	0.059	2.000	0.930	1,073.0	9.3E-07	89.1	3.07
0.647	0.288	0.007	0.059	2.000	0.930	1,182.0	8.7E-07	90.5	2.91
0.647	0.288	0.007	0.059	2.000	0.930	1,281.0	7.8E-07	92.3	2.67
0.647	0.288	0.007	0.059	2.000	0.930	1,377.0	7.3E-07	94.8	2.55
0.647	0.288	0.007	0.059	2.000	0.930	1,485.0	6.9E-07	99.4	2.53
0.647	0.288	0.007	0.059	2.000	0.930	1,281.0	8.1E-07	92.3	2.76
0.647	0.288	0.007	0.059	2.000	0.930	1,078.0	9.6E-07	89.2	3.19
0.647	0.288	0.007	0.059	2.000	0.930	874.0	1.2E-06	86.4	3.91
0.647	0.288	0.007	0.059	2.000	0.930	1,492.0	6.9E-07	99.8	2.54
0.647	0.288	0.007	0.059	2.000	0.930	1,581.0	6.8E-07	105.8	2.64
0.647	0.288	0.007	0.059	2.000	0.930	1,680.0	6.5E-07	115.3	2.75
0.647	0.288	0.007	0.059	2.000	0.930	1,781.0	6.3E-07	128.6	2.95
0.683	0.294	0.023	-	1.916	0.928	992.9	6.0E-07	85.8	1.89
0.683	0.294	0.023	-	1.916	0.928	1,365.9	5.0E-07	91.9	1.67
0.683	0.294	0.023	-	1.916	0.928	1,552.7	5.0E-07	101.3	1.82
0.683	0.294	0.023	-	1.916	0.928	2,074.9	5.0E-07	165.0	2.91
0.683	0.295	0.022	-	1.946	0.931	1,067.3	6.5E-07	87.6	2.12
0.700	0.295	0.022	-	1.946	0.931	1,244.0	5.9E-07	90.7	1.94
0.700	0.295	0.022	-	1.946	0.931	1,432.2	5.7E-07	96.2	1.96
0.700	0.295	0.022	-	1.946	0.931	1,625.2	5.4E-07	109.3	2.11
0.700	0.295	0.022	-	1.946	0.931	1,821.1	5.3E-07	135.4	2.53
0.700	0.295	0.022	-	1.946	0.931	2,020.0	5.2E-07	166.3	3.03
0.700	0.295	0.022	-	1.946	0.931	2,192.5	5.3E-07	149.1	2.74
0.683	0.295	0.022	-	1.932	0.930	1,064.7	5.9E-07	87.1	1.91
0.683	0.295	0.022	-	1.932	0.930	1,244.0	5.6E-07	89.6	1.83
0.683	0.295	0.022	-	1.932	0.930	1,432.5	5.1E-07	95.0	1.77
0.683	0.295	0.022	-	1.932	0.930	1,625.2	4.9E-07	108.0	1.92
0.683	0.295	0.022	-	1.932	0.930	1,821.5	4.9E-07	134.0	2.35
0.683	0.295	0.022	-	1.932	0.930	2,020.0	4.9E-07	164.8	2.86
0.683	0.295	0.022	-	1.932	0.930	2,191.5	5.0E-07	147.6	2.62
0.682	0.294	0.023	-	1.926	0.934	1,062.8	5.9E-07	86.9	1.91
0.682	0.294	0.023	-	1.926	0.934	1,245.0	5.4E-07	89.4	1.79
0.682	0.294	0.023	-	1.926	0.934	1,434.8	5.1E-07	94.9	1.77
0.682	0.294	0.023	-	1.926	0.934	1,627.1	5.1E-07	107.9	2.00
0.682	0.294	0.023	-	1.926	0.934	1,822.1	5.0E-07	133.8	2.42
0.682	0.294	0.023	-	1.926	0.934	2,020.0	5.0E-07	164.5	2.94
0.682	0.294	0.023	-	1.926	0.934	2,191.5	5.1E-07	147.7	2.66
0.683	0.295	0.022	-	2.000	0.936	872.0	1.2E-06	86.5	3.86

(*Continued*)

TABLE 6 (*Continued*)
Thermal Diffusivity

U	Pu	Am	Np	O/M	ρ (%TD)	Temp. (K)	k (m²/s)	C_p^a (J/molK)	λ^a (W/mK)
0.683	0.295	0.022	–	2.000	0.936	975.0	1.0E-06	88.1	3.46
0.683	0.295	0.022	–	2.000	0.936	1,074.0	9.5E-07	89.4	3.18
0.683	0.295	0.022	–	2.000	0.936	1,174.0	8.7E-07	90.7	2.94
0.683	0.295	0.022	–	2.000	0.936	1,279.0	8.1E-07	92.5	2.79
0.683	0.295	0.022	–	2.000	0.936	1,372.0	7.6E-07	94.9	2.67
0.683	0.295	0.022	–	2.000	0.936	1,474.0	7.2E-07	99.3	2.65
0.683	0.295	0.022	–	2.000	0.936	1,573.0	6.7E-07	105.7	2.60
0.683	0.295	0.022	–	2.000	0.936	1,674.0	6.4E-07	115.4	2.71
0.683	0.295	0.022	–	2.000	0.936	1,774.0	6.1E-07	128.8	2.89
0.592	0.400	0.008	–	2.000	0.908	980.0	1.0E-06	88.7	3.28
0.592	0.400	0.008	–	2.000	0.908	1,082.0	9.1E-07	90.1	3.01
0.592	0.400	0.008	–	2.000	0.908	1,167.0	8.3E-07	91.4	2.77
0.592	0.400	0.008	–	2.000	0.908	1,273.0	7.7E-07	93.7	2.62
0.592	0.400	0.008	–	2.000	0.908	1,367.0	7.2E-07	97.2	2.54
0.592	0.400	0.008	–	2.000	0.908	1,469.0	6.8E-07	103.4	2.55
0.592	0.400	0.008	–	2.000	0.908	1,567.0	6.3E-07	113.0	2.55
0.592	0.400	0.008	–	2.000	0.908	1,668.0	6.0E-07	127.3	2.72
0.592	0.400	0.008	–	2.000	0.908	1,767.0	5.9E-07	145.5	3.06
0.787	0.200	0.013	–	2.000	0.940	996.0	1.1E-06	88.1	3.50
0.787	0.200	0.013	–	2.000	0.940	1,094.0	9.5E-07	89.2	3.17
0.787	0.200	0.013	–	2.000	0.940	1,177.0	8.8E-07	90.2	2.98
0.787	0.200	0.013	–	2.000	0.940	1,282.0	8.0E-07	91.6	2.75
0.787	0.200	0.013	–	2.000	0.940	1,375.0	7.5E-07	93.4	2.61
0.787	0.200	0.013	–	2.000	0.940	1,477.0	7.0E-07	96.3	2.50
0.787	0.200	0.013	–	2.000	0.940	1,575.0	6.9E-07	100.5	2.55
0.787	0.200	0.013	–	2.000	0.940	1,666.0	6.6E-07	106.0	2.56
0.787	0.200	0.013	–	2.000	0.940	1764	6.315E-07	114.1	2.64
0.6	0.300	0.100	–	2.000	0.8678	286	1.75E-06	70.1	4.40
0.6	0.300	0.100	–	2.000	0.8678	383	1.44E-06	73.7	3.80
0.6	0.300	0.100	–	2.000	0.8678	476	1.23E-06	76.5	3.36
0.6	0.300	0.100	–	2.000	0.8678	571	1.10E-06	79.2	3.10
0.6	0.300	0.100	–	2.000	0.8678	676	9.85E-07	81.9	2.86
0.6	0.300	0.100	–	2.000	0.8678	774	9.06E-07	84.2	2.69
0.6	0.300	0.100	–	2.000	0.8678	871	8.52E-07	85.9	2.58
0.6	0.300	0.100	–	2.000	0.8678	973	7.80E-07	87.4	2.39
0.6	0.300	0.100	–	2.000	0.8678	1074	7.26E-07	88.7	2.25

[a] Cp and λ were calculated by Eqs. (3.70) and (3.79), respectively, in this work.

TABLE 7
Oxygen Diffusion Coefficients

U	Pu	O/M Ratio Initial	O/M Ratio Final	Temperature (K)	Self- (m²/s)	Chemical (m²/s)
0.80	0.20	2.000	1.994	1,523		2.9E-10
0.80	0.20	2.000	1.992	1,573		2.9E-10
0.80	0.20	1.994	1.992	1,573		3.2E-10
0.80	0.20	1.996	1.993	1,573		2.9E-10
0.80	0.20	2.000	1.992	1,573		3.3E-10
0.80	0.20	2.000	1.997	1,573		3.2E-10
0.80	0.20	2.000	1.994	1,573		3.3E-10
0.80	0.20	2.000	1.992	1,573		3.0E-10
0.80	0.20	2.000	1.990	1,623		3.2E-10
0.80	0.20	1.999	1.990	1,623		3.8E-10
0.80	0.20	1.999	1.998	1,623		2.6E-10
0.80	0.20	2.000	1.999	1,623		3.4E-10
0.80	0.20	1.998	1.997	1,623		2.6E-10
0.80	0.20	1.999	1.995	1,623		3.6E-10
0.70	0.30	2.000	1.992	1,523		1.4E-10
0.70	0.30	2.000	1.992	1,523		1.5E-10
0.70	0.30	2.000	1.985	1,573		1.2E-10
0.70	0.30	2.000	1.987	1,573		1.6E-10
0.70	0.30	2.000	1.991	1,573		1.8E-10
0.70	0.30	2.000	1.989	1,573		1.8E-10
0.70	0.30	1.990	1.987	1,573		1.6E-10
0.70	0.30	2.000	1.979	1,623		2.5E-10
0.70	0.30	2.000	1.988	1,623		2.0E-10
0.70	0.30	1.988	1.984	1,623		2.2E-10
0.00	1.00	2.000	1.984	1,473		4.4E-11
0.00	1.00	2.000	1.982	1,673		4.8E-11
0.00	1.00	2.000	1.958	1,673		6.8E-11
0.00	1.00	2.000	1.995	1,773		7.8E-11
0.00	1.00	2.000	1.989	1,773		6.2E-11
0.00	1.00	2.000	1.984	1,773		6.2E-11
0.00	1.00	2.000	1.982	1,773		5.0E-11
0.00	1.00	2.000	1.980	1,773		4.2E-11
0.00	1.00	2.000	1.970	1,773		4.5E-11
0.00	1.00	2.000	1.956	1,773		7.5E-11
0.00	1.00	2.000	1.981	1,873		5.2E-11
0.00	1.00	2.000	1.946	1,873		7.0E-11
0.80	0.20	2.000	1.999	1,473		8.9E-10
0.80	0.20	2.000	1.998	1,473		5.3E-10
0.80	0.20	2.000	1.999	1,573		5.7E-10
0.80	0.20	2.000	1.997	1,573		1.0E-09
0.80	0.20	2.000	1.996	1,573		5.6E-10
0.80	0.20	2.000	1.996	1,673		9.6E-10
0.80	0.20	2.000	1.995	1,673		4.8E-10

(Continued)

TABLE 7 (*Continued*)
Oxygen Diffusion Coefficients

U	Pu	O/M Ratio Initial	O/M Ratio Final	Temperature (K)	Self- (m²/s)	Chemical (m²/s)
0.80	0.20	2.000	1.999	1,673		1.7E-09
0.80	0.20	2.000	1.991	1,773		1.1E-09
0.80	0.20	2.000	1.996	1,773		1.7E-09
0.80	0.20	2.000	1.991	1,773		2.0E-09
0.80	0.20	2.000	1.992	1,773		1.3E-09
0.80	0.20	2.000	1.994	1,873		2.1E-09
0.80	0.20	1.999	1.983	1,873		5.6E-10
0.80	0.20	2.000	1.986	1,873		1.1E-09
0.80	0.20	1.999	1.984	1,873		8.7E-10
0.70	0.30	2.000	1.999	1,473		5.3E-10
0.70	0.30	2.000	1.997	1,573		5.6E-10
0.70	0.30	2.000	1.996	1,573		2.4E-10
0.70	0.30	1.999	1.993	1,673		3.5E-10
0.70	0.30	1.999	1.994	1,673		2.2E-10
0.70	0.30	1.999	1.996	1,673		7.2E-10
0.70	0.30	1.999	1.992	1,773		9.1E-10
0.70	0.30	1.999	1.989	1,773		5.8E-10
0.70	0.30	1.999	1.995	1,773		3.7E-10
0.70	0.30	1.999	1.982	1,873		1.6E-09
0.70	0.30	1.999	1.992	1,873		8.4E-10
0.70	0.30	1.999	1.992	1,873		6.0E-10
0.80	0.20	2.000		1,673	1.40E-11	
0.80	0.20	2.001		1,673	5.31E-12	
0.80	0.20	2.000		1,773	7.95E-12	
0.80	0.20	2.001		1,773	1.25E-11	
0.80	0.20	2.000		1,873	2.12E-11	
0.80	0.20	2.000		1,873	1.45E-11	
0.80	0.20	2.003		1,873	5.56E-11	
0.80	0.20	2.001		1,873	5.15E-11	
0.70	0.30	2.000		1,673	1.40E-12	
0.70	0.30	2.002		1,673	7.17E-12	
0.70	0.30	2.001		1,673	9.81E-12	
0.70	0.30	1.997		1,673	5.10E-12	
0.70	0.30	2.000		1,773	2.00E-11	
0.70	0.30	2.003		1,773	1.59E-11	
0.70	0.30	2.001		1,773	7.46E-12	
0.70	0.30	1.997		1,773	4.74E-11	
0.70	0.30	2.000		1,873	3.14E-11	
0.70	0.30	2.000		1,873	2.35E-11	
0.70	0.30	2.000		1,873	1.92E-11	
0.70	0.30	1.997		1,873	5.66E-11	
0.70	0.30	2.003		1,873	4.24E-11	
0.70	0.30	2.003		1,873	3.13E-11	
0.70	0.30	2.001		1,873	1.15E-11	

Index

For Product Safety Concerns and Information please contact our EU
representative GPSR@taylorandfrancis.com
Taylor & Francis Verlag GmbH, Kaufingerstraße 24, 80331 München, Germany